DIMORPHIC FUNGI: THEIR IMPORTANCE AS MODELS FOR DIFFERENTIATION AND FUNGAL PATHOGENESIS

Edited By

José Ruiz-Herrera

Unidad Irapuato
Centro de Investigación y de Estudios Avanzados del Instituto
Politécnico Nacional
Irapuato, Gto.
México

eBooks End User License Agreement

DEDICATION

This eBook is dedicated to all the members of my family for their understanding, their encouragement of my work, their help, and all the happiness they have brought to my life.

Special thanks are given to the authors of each one of the chapters of the eBook. Were not for their outstanding work and their unselfish collaboration, this volume would had never been written.

Additionally, I thank all my collaborators and students for their help in the research carried out in my laboratory, some of whose data appear in this volume.

CONTENTS

CHAPTERS

FOREWORD

"Fungal dimorphism" refers to the ability of some fungal species to grow under two morphological forms, feature that has fascinated the scientists for almost a Century. The original interest in fungal dimorphism was associated to pathogenesis but it is now clear that it implies something much more important than the exclusive study of the final morphology of the fungal cells. The original definition overlooks the importance of dimorphism as a process of cellular differentiation. We can inquire which mechanisms are involved in the morphological changes and the possibilities are that certain genes are expressed only in a specific morphology or that the gene products are assembled in the cell wall in a different way to give an alternative molecular architecture. Only an in deep understanding of the processes involved in dimorphism will be possible with development of a wide-ranging list of experimental procedures combined together with the new genetic techniques, the rapid access to genomic information and bioinformatic analyses of many of these fungal species. It can be said that the new information collected literally forged a new frontier of research for understanding the molecular nature and the control of the fungal dimorphism. The aim of this eBook is to present in a single volume a review as current as possible of dimorphism among fungi. It addresses the actual state of art on relevant aspects of dimorphism in a few fungal species by different authors as it is increasingly clear that reviewing this subject adequately is probably beyond the capability of a single author. The text is organized in a way that attempts to enhance readability and should be of interest not only to those doing research in this specific area (including detection of potential targets for development of new antifungal drugs), but also to those interested in other aspects of fungal cells and for those in general biology.

Rafael Sentandreu
Facultat de Farmacia
Universitat de Valencia
España

PREFACE

Dimorphism is the property of different fungal species to grow as budding yeasts or mycelium depending on the environmental conditions. Dimorphism involves extensive changes in cell physiology and morphology in response to external signals, making this phenomenon a superb model for the study of the differentiation processes occurring in eukaryotic organisms. Additionally, it must be pointed out that the most important human and some plant pathogenic fungi are dimorphic, displaying different morphologies when growing as saprophytes, or inside their hosts, suggesting that this process may be an ideal target for the design of efficient antimycotic drugs.

Much information of the molecular bases of the dimorphic transition in fungi has been generated in recent years, and the important advances in this important subject in the most widely studied fungal species are presented and discussed in this volume by different specialists with a well gained reputation in the field.

I am certain that the contents of this eBook will be an important source of information to all the scholars working or interested in human, animal and plant pathogenesis; fungal genetics, molecular biology and development; evolution, and cell differentiation.

José Ruiz-Herrera
Irapuato Gto.
Mexico

List of Contributors

Campos-Góngora, E.

Departamento de Ingeniería Genética, Unidad Irapuato, Centro de Investigación y de Estudios Avanzados del IPN[1]; and Centro de Investigación en Nutrición y Salud Pública, Universidad Autónoma de Nuevo León[2], Monterrey, NL México. Km. 9.6 Libramiento Norte Carretera Irapuato-León. CP. 36500 Apartado Postal 629; Tél: +5246239600; Fax: +524626235948; E-mail. Educampos@hotmail.com

Calabrese, E.

Dept. of Pharmaceutical and Biomedical Science, University of Salerno, Italy. E-mail: bmaresca@unisa.it

Deshpande, MV.

Biochemical Sciences Division, National Chemical Laboratory, Pune 411008, India. E-mail: mv.deshpande@ncl.res.in

Ghormade, V.

Centre for Nanobioscience, Agharkar Research Institute, Pune 411004, India. E-mail: mv.deshpande@ncl.res.in

Granata, I.

Dept. of Pharmaceutical and Biomedical Science, University of Salerno, Italy. Email: bmaresca@unisa.it

Guerrero-González, M de la L.

División de Biología Molecular, Instituto Potosino de Investigación Cientifica y Tecnológica (IPICYT), San Luis Potosí, México, Camino a la Presa de San José 2055, Apartado Postal 3-74 Tangamanga, C.P. 78216, San Luis Potosí, SLP, México. E-mail: maria.gonzalez@ipicyt.edu.mx

Jiménez-Bremont, JF.

División de Biología Molecular, Instituto Potosino de Investigación Cientifica y Tecnológica (IPICYT), San Luis Potosí, México, Camino a la Presa de San José 2055, Apartado Postal 3-74 Tangamanga, C.P. 78216, San Luis Potosí, SLP, México. E-mail: jbremont@ipicyt.edu.mx

León-Ramírez, CG.

Departamento de Ingeniería Genética, Unidad Irapuato. Centro de Investigación y de Estudios Avanzados del IPN, México. Centro de Investigación y de Estudios Avanzados del Instituto Politécnico Nacional. Km. 9.6 Libramiento Norte Carretera Irapuato-León. CP. 36500 Apartado Postal 629; Tél: +5246239600; Fax: +524626235948; E-mail: cleon@ira.cinvestav.mx

Lopes-Bezerra, LM.

Laboratory of Cellular Mycology and Proteomics, Biology Institute, State University of Rio de Janeiro (UERJ), Rua São Francisco Xavier 524 PHLC, 20550-013, Rio de Janeiro, Brazil. E-mail: leila.lopes_bezerra@pq.cnpq.br

Maresca, B.

Dept. of Pharmaceutical and Biomedical Science, University of Salerno, Italy. E-mail: bmaresca@unisa.it

Nascimento, RC.

Laboratory of Cellular Mycology and Proteomics, Biology Institute, State University of Rio de Janeiro (UERJ), Rua São Francisco Xavier 524 PHLC, 20550-013, Rio de Janeiro, Brazil. E-mail: rosana.cicera@gmail.com

Niño-Vega, G.

Laboratorio de Micología, Centro de Microbiología y Biología Celular, Instituto Venezolano de Investigaciones Científicas (IVIC), Apartado 20632, Caracas 1020A, Venezuela. E-mail: sanblasg@ivic.gob.ve

Pathan, E.

Biochemical Sciences Division, National Chemical Laboratory, Pune 411008, India. E-mail: mv.deshpande@ncl.res.in

Rodríguez-Hernández, AA.

División de Biología Molecular, Instituto Potosino de Investigación Cientifica y Tecnológica (IPICYT), San Luis Potosí, México, Camino a la Presa de San José 2055, Apartado Postal 3-74 Tangamanga, C.P. 78216, San Luis Potosí, SLP, México. E-mail: aida.rodriguez@ipicyt.edu.mx

Rodríguez-Kessler, M.

Facultad de Ciencias, Universidad Autónoma de San Luis Potosí, México. Av. Salvador Nava s/n, Zona Universitaria C.P. 78290, San Luis Potosí, SLP, México. E-mail:mrodriguez@uaslp.mx

Ruiz-Herrera, J.

Departamento de Ingeniería Genética, Unidad Irapuato. Centro de Investigación y de Estudios Avanzados del IPN, México. Km. 9.6 Libramiento Norte Carretera Irapuato-León. CP. 36500 Apartado Postal 629; Tél: +5246239600; Fax: +524626235948; E-mail: jruiz@ira.cinvestav.mx

San-Blas, G.

Laboratorio de Micología, Centro de Microbiología y Biología Celular, Instituto Venezolano de Investigaciones Científicas (IVIC), Apartado 20632, Caracas 1020A, Venezuela. E-mail sanblasg@ivic.gob.ve

2

CHAPTER 1

An Introduction to Fungal Dimorphism

José Ruiz-Herrera[1,*] and Eduardo Campos-Góngora[1,2]

[1]Departamento de Ingeniería Genética, Unidad Irapuato, Centro de Investigación y de Estudios Avanzados del IPN, Mexico and [2]Centro de Investigación en Nutrición y Salud Pública, Universidad Autónoma de Nuevo León, Monterrey, NL México

Abstract: The members of some fungal species have the ability to grow in the form of yeasts or mycelium depending on the environmental and some internal conditions. This phenomenon, denominated "dimorphism" is not exclusive of some fungal taxa, but examples represent almost all groups. Fungal dimorphism is an important phenomenon from both applied and basic concepts. In the former aspect because a significant number of species pathogenic for plants, animals, and specially humans are dimorphic, the causal agents displaying different forms during their saprophytic and pathogenic stages, and more specifically because it has been demonstrated that inhibition of the dimorphic transition by drugs or mutation blocks the pathogenic development. Regarding the second aspect, the dimorphic transition has all the elements to consider it as an example of cell differentiation, and as such it constitutes a basic model for the study of this important biological phenomenon. In this chapter we analyze dimorphism on this line and propose its study through its division into four blocks of reactions: stimulus reception, translation of the stimulus, change in the developmental program and final outcome

Keywords: Dimorphism, fungi, yeast, mycelium, hyphae, pseudomycelium, cell differentiation, development, morphogenesis, molecular taxonomy, phylogeny, morphology index, virulence, cell wall.

INTRODUCTION

Dimorphism is the ability of fungi to grow in the form of yeast-like cells that multiply through budding, or in the form of hyphae characterized by apical growth, depending on the environmental conditions or in response to internal stimuli.

Dimorphism and the ability to grow as unicellular yeasts or multicellular filamentous forms permit fungi to adapt to different conditions, or develop distinct morphologies depending on their development status. In Nature, this phenomenon is employed by fungi, not only as a way to adapt to new prevailing conditions or develop specific functions during their life cycle, but also as a way to ensure infection of a host as occurs with pathogenic fungi.

This phenomenon, originally described and correctly interpreted by Pasteur, against the concept of species variation, led to the use of the term to describe the phenotypic duality of forms in the distinct phases of some parasitic fungi, and in a more general use to the existence of a saprophytic phase (normally hyphae) and a parasitic phase (normally yeast-like), [1] but other terms have emerged in recent years, namely the absurd term of "Morphological Dimorphism" to refer to the ability to switch between unicellular yeast cells and filamentous forms either hyphae and pseudohyphae [2]. Nevertheless, taken in its simplest form, the term dimorphism has been applied, as indicated above to the ability of fungi to differentiate in two different forms during their life cycle. One medical importance of dimorphism is, as indicated above, that many fungi pathogenic for the human are dimorphic: *e.g. Blastomyces dermatitidis, Paracoccidioides brasiliensis, Sporothrix schenckii, Exophiala werneckii, Exopliala (Wangiella) dermatitidis, Coccidioidis immitis, Emmonsia parva, Phialophora verrucosa, Candida albicans, Mucor spp., Histoplasma capsulatum.* Table **1**.

---
*Address correspondence José Ruiz Herrera: Departamento de Ingeniería Genética, Unidad Irapuato. Centro de Investigación y de Estudios Avanzados del IPN, México; Tél: +5246239600; Fax: +524626235948; E-mail: jruiz@ira.cinvestav.mx

Table 1: Examples of dimorphic fungi

SPECIES	EFFECTOR	REQUIREMENT FOR MYCELIUM FORMATION
Candida albicans	Change in temperature	Higher
	pH	Neutral
	C source	GlcNAc
	Other	Serum
Yarrowia lipolytica	pH	Neutral
	Buffer	Citrate
	Other	Serum
Ustilago maydis	pH	Acid
	C source	Fatty acids
Paracoccidioides brasiliensis	Change in temperature	Lower
Coccidioides immitis	Change in temperature	Lower
Blastomyces dermatitidis	Change in temperature	Lower
Histoplasma capsulatum	Change in temperature	Lower
Mucor rouxii	Oxygen	High tension
	C source	Non-fermentable
Mucor racemosus	Atmosphere	Aerobic
	C. source	Non-fermentable

These pathogenic species grow in general in a mycelial form in their saprophytic stage, or yeast-like form in their parasitic phase during host invasion, although, as expected, there are exceptions, such as *C. albicans* whose saprophytic form is yeast-like, and the invasive form is mycelial. When fungi are able to adopt more than two distinct morphologies, they are called polymorphic, an example being the same *C. albicans*.

In recent years, the development of new analytical methods, tools for genetic manipulation, and availability of sequenced genomes, among others, has allowed significant progress in understanding the regulation of the processes implicated in fungal dimorphism. Now, we know that during the dimorphic transition, fungi change not only in shape, but also in a plethora of traits, such as metabolic activities, the composition and structure of the cell wall, the formation of antigenic molecules, and in the case of pathogenic fungi, the expression of virulence traits [3]. The occurrence of these morphological and physiological changes taking place during the dimorphic transition, lead to the always unanswered question: cause or effect?.

The environmental conditions that induce the dimorphic switch are mainly, changes in temperature, pH, nutrients (such as carbon or nitrogen sources), presence of specific compounds, including those produced by susceptible hosts, *etc*. It has been said that in general, lower temperatures, acidic pH or high concentrations of glucose stimulate the production of unicellular forms (reviewed in [2]), but this can change depending of each species in particular. Examples of internal conditions that during the life cycle induce morphogenetic factors that end in a dimorphic switch are multiple. As an example of this, the haploid, saprophytic stage of *Ustilago maydis* is yeast-like, whereas the dikaryotic virulent stage of the fungus is mycelial.

As a word of caution it must be stressed that the normal morphological changes that occur during the life cycle of most fungi, *e.g.* the formation of spherical or odd-looking spores by mycelial fungi, do not enter into the dimorphic phenomenon *sensu stricto*. It must be recalled that the simplest definition of a yeast is a cell that reproduces by budding.

Brief Historic Account of the Study of Dimorphism

Dimorphism was originally described in *Mucor* species, as the first example of the capacity of some fungal species to grow in hyphal or yeast-like forms depending on the environmental conditions [4]. Originally considered an example of the transmutations of microbial species: a yeast and a mold [5, 6], it was recognized by Pasteur as a case of a morphogenetic dichotomy induced by the environmental conditions [7]. According to his results and interpretations, *Mucor* behaved as a respiring mold under aerobic conditions, and not only changed its metabolism, but also its morphology, to behave as fermenting yeast, producing alcohol and carbon dioxide in anaerobiosis. Although he suggested that CO_2 tension was perhaps more important than anaerobiosis for yeast-like development, it was Brefeld [8] who conclusively demonstrated this, when he obtained hyphal growth of *M. racemosus* under a H_2 atmosphere. Finally, Wehmer [9] was able to separate morphology from physiological behavior, when he demonstrated that yeasts and mycelium obtained in anaerobiosis produced similar amounts of ethanol by sugar fermentation. More recently the studies [10] corroborated beyond any doubt that that CO_2, and not the mere absence of oxygen was the morphogenetic factor that induced yeast-like growth in *M. rouxii*. The *Mucorales* group is an example that related species may be or not dimorphic, and that the conditions that induce the dimorphic switch may be different. Thus, *Rhizopus* is not dimorphic, and *Mucor bacilliformis* does not require a high tension of CO_2 to grow in the yeast form, it only requires an anaerobic environment.

Fungal Dimorphism as a Model for Cell Differentiation

There exist a large number of definitions for cell differentiation; among them we may cite the following: Nickerson and Bartnicki-García [11] made the following definitions: "*growth* is defined as an increase in cellular mass (dry wt); *development*, which should not be used as a synonym of growth, refers to the sequence of structural and functional changes occurring during the life cycle of an organism; d*ifferentiation* connotes the acquisition of a determined function or structure during development and, although commonly used as a synonym of development, is only a component thereof. *Morphogenesis* encompasses those aspects of development related to morphological changes". Garrod and Ashworth [12] established that ". development in multicellular organims may be usefully, if somewhat arbitrarily, divided into three separate aspects: 1. ".differentiation involving the structural and functional specialization of individual cells from one of a number of common basic stem cells which are usually competent to differentiate in several different way. Differentiation may be regarded as essentially an intracellular process." 2. ".Pattern formation, is concerned with the spatial organization of differentiation: that is how cells of different types develop in the correct spatial, temporal and proportional relationships to each other" 3. ".Morphogenesis refers to the development of the shape and form of the organism and its individual parts.". Harold has stated that "growth, morphogenesis and development thus appear to be fundamentally problems of biological order."; and then added "the term morphogenesis will refer to the processes that generate the forms of cells in the course of their growth, division or development" [13].

Wright made a stronger emphasis in the time factor, and concluded that "biochemical differentiation is a process of cellular specialization in which a substance is accumulated to a unique extent over a specified period of time" [14].

But a more simple definition that we coined a long time ago can be applied here: "Differentiation is the sum of events that correctly applied in time and space result in the specialization of a cell without alterations in its genetic characteristics" [15]. It may be seen that dimorphism perfectly fits into this definition, with the only exception that some authors consider that differentiation is an irreversible phenomenon, and dimorphic transition is not. However, we think that this property is an advantage in the experimental analysis of the phenomenon. This characteristic, the fact that it is possible to obtain perfectly viable mutants that have lost their dimorphic capacity (*i.e.*, they are monomorphic) makes dimorphism an excellent model in the study of cell differentiation.

In general terms we may represent a differentiation phenomenon, trying to gather the ideas just presented, as the general scheme shown in Fig. **1a**) According to the scheme, starting from a tutipotent and undiffetiated cell (we beg forgiveness for introducing part of the defined phenomenon into the definition), such as an egg cell, a spore, *etc.*, different types of cells are going to be formed. In the scheme this is simply represented by a change in the geometry of the figures. In some systems differentiation ends here, with the production of several types of cells differing in form and/or function. But in other systems, the stem cells give rise to other cell lines with different characteristics. This further specialization may be sequential until the end of the differentiative pathway, *i.e.* cells that cannot give rise themselves to a different line of cells, are formed. Normally the cells at each differentiative step cannot give rise to a cell line coming from a different pathway. But this characteristic may depend on the organism and the conditions under which the process is taking place.

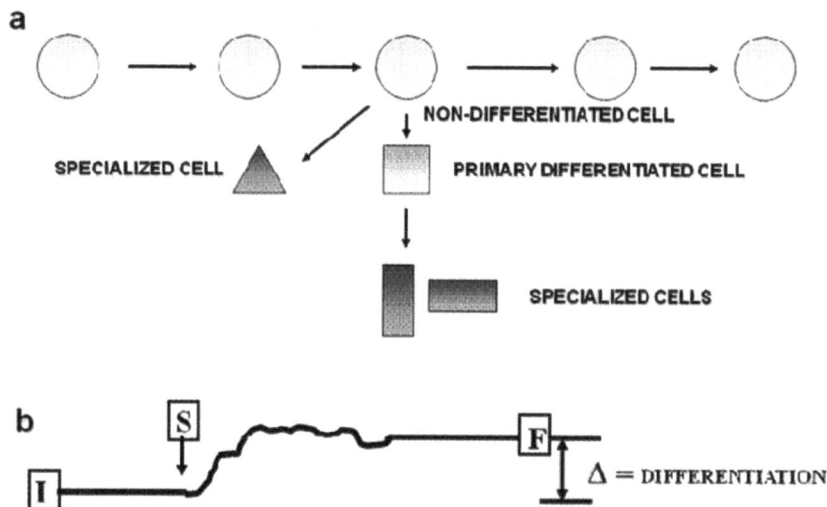

Figure 1: Schematic representation of differentiation. a. Differentiation involves processes that do not have any effect on the reproductive mechanisms of the cells, since they do not involve modifications of the nuclear DNA. Differentiation may give rise to specialized cells either directly as occurs in fungi or directly and indirectly as occurs in higher eukaryotes. b. A mechanistic representation of differentiation. A non-differentiated cell that grows in steady state conditions (I) upon reception of a determined stimulus (S), enters into a state of non-equilibrium, until reaching a different level of steady state (F). The difference between F and I stages corresponds to the differentiation process.

The problem acquires a further order of complexity when we attempt to establish *a priori* how different must the cells be from their immediate precursor in order to consider that they have suffered a differentiation process. According to the concept of Wright [14] (see above), it is only necessary that it accumulates a specific substance at some time. But in order to exemplify how tricky this may be, let us take the example of *Escherichia coli,* which when grown in the absence of lactose (or any inducer in general) has only a low amount of the enzyme that can hydrolyze β-galactosides. But if grown in the presence of an inducer, accumulates huge levels of the hydrolytic enzyme. Can we say that the bacterium has become differentiated? It becomes obvious that under these premises, the phenomenon, which originally we had judged to be qualitative, has become a quantitative one.

In order to cope with this difficulty, we may recourse to an operational definition which is represented in Fig. **1b**. According to the scheme, an undifferentiated cell or organism maintains itself in a resting steady state conditions. If at certain moment it receives a specific stimulus, it abandons the steady state and loses its original equilibrium, until it reaches the steady state again, but at a different level from the original one. According to this concept, differentiation would be the difference between both steady state levels. It is not to be expected that the magnitude of this difference satisfies all the criteria, and it is not necessary it does, since, coming back again to the reasoning from Barbara Wright [14], we must not be interested in the definition, but in the phenomenon.

If the interest behind these analyses is to understand the mechanisms involved in the operation of the stimulation process, or how the cells or organisms achieve their new steady state, *etc.*, the magnitude of the change becomes irrelevant. The important result will be the elaboration of schemes that satisfy our knowledge of the *operation* of the phenomenon. As Herrman and Toole pointed at a long time ago, "the absence of a "general theory" of differentiation may be of less importance (to biology)... than the complete analysis of the molecular mechanisms of specific developmental processes, a worthwhile end in itself" [16].

In connection with these ideas, the next problem to be discussed would be, Is it *a priori* acceptable that there is a cause-effect relationship during differentiation?, and there is a single cause and a single effect during differentiation?. This is a question difficult to answer, since sometimes it is extremely difficult to decide which is the cause and which the effect in many of these phenomena or whether an organism that receives a single stimulus, displays the same response, or if different stimuli induce the same effect. A further problem is whether we should consider that the chain of events triggered by a stimulus operates in a linear fashion. We may anticipate that this is rarely the *modus operandi* of differentiation mechanisms. Although it is possible to demonstrate the existence of some unidirectional chains of events, it is more general that differentiative phenomena involve oscillatory changes in the systems that maintain the cellular steady state. These depend themselves on the interaction of metabolites and enzymes, whose levels change in response to alteration in the rates of transcription, translation, and post-translational modifications. These changes lead to a complex network of interactions, rather than to fixed unidirectional changes. Finally, an important question in the analysis of differentiative phenomena is whether there exist elements, reactions, or components that are common to the several differentiative processes displayed by different organisms. In the case of the dimorphism problem, are there common pathways in the dimorphic switch of the different dimorphic fungi? A point purposely left at the end of these considerations regards to the level at which differentiation should be analyzed. Although we are dealing with a biological problem, we must accept that biology involves different levels of study. It must be accepted in advance that to obtain a global explanation of differentiation manipulating single concepts of physics, chemistry, biochemistry, molecular biology or classical genetics, *etc.*, given the extreme complexity of these phenomena, it is necessary to include all these concepts into a further level of organization in their deciphering. As Peter Mitchell stated [17] "the sum of non scalar reactions can not produce a vectorial phenomenon". It appears logical that we cannot understand differentiation by summing up all the isolated biochemical reactions; that is, the more detailed knowledge we reach of the isolated reactions, the less holistic view we have of the problem (according to the popular dictum, the trees do not let us see the forest). As a tactical maneuver, we may advocate an approach that simplifies the problem. This approach is to make a simplistic division of the process into blocks of information which aprioristically (although probably incorrectly) we have assumed to follow a vectorial direction. We accept beforehand that this approach is reductionist, but we consider that if we can separate and analyze full sets of the reactions, we have a better chance to solve the puzzle, and obtain a general scheme of the phenomenon in the long run. Division of the phenomena into four steps might be the following (Fig. **2**):

1. Reception of the stimulus. Considering that differentiation processes are triggered by different stimuli, we may anticipate that there must be adequate receptors that perceive the corresponding stimuli.

2. Decoding of the stimulus. This is the process that transforms the information brought in by the stimulus into a "language" that the cell or the organism "understands". In order that this occurs it is necessary that the excitation traverses through a chain of reactions, generally known as signal transduction pathways.

3. Change in program. Through its operation, the information received will change the established steady state condition of the cell or organism at the transcriptional, translational, and post-translational levels, giving rise to the establishment of a different program, and a new steady state condition.

4. Final reaction. This would be the expression of changes in the physiology or/and morphology in a quantitative and qualitative fashion. Alterations in the structure of the cell or the organism depends on vectorial processes, in contrast with the other phenomena, which are mostly scalar. In fungi, and specifically in dimorphism, the morphogenetic alterations that result from this program operate on the cell wall, which is responsible for cell shape.

When analyzed under this optic it is difficult to escape ourselves from comparing those processes to the Aristotelic concept followed by Leibnitz, and taken to the extreme by Driesch of entelechy, "a real thing that has in itself the principle of action that leads by itself to its own purpose". It is easy to see why the teleological assumption that all organisms, even the simplest ones, appear to lead by themselves to their purpose, seems even natural.

Perception of Stimuli

General Concepts

According to the ideas discussed above, we may conclude that fungi are subjected to complex differentiation phenomena (I avoid the common expression "are able to carry out.", to stress the non-finalist idea behind the phenomena). Among these phenomena, dimorphism is an important one as stressed before. Dimorphic transition in fungi occurs in response to stimuli from the environment, some of which may be extremely specific, or by internal signals during their life cycle. An interesting observation is that the same response may be obtained by different stimuli, for example use of GlcNAc as alternative carbon source or addition of serum, both induces mycelial growth in *C. albicans* [18, 19]. Alternatively, the same stimulus may give rise to contrary effects in different fungi. This is the case for heat shock which induces the mycelium-to-yeast dimorphic transsition in a number of fungi pathogenic for man, whereas on the other hand, it induces mycelial growth in *C. albicans*. Similarly, growth at acid or alkaline pH may give different results, thus acid pH prevents mycelial growth of *C. albicans* [20, 21] whilst it induces mycelial growth in *U. maydis* [22, 23]. It may be anticipated that all these specific, general and contradictory responses to different stimuli depend in a first degree on the existence of specific receptors for them, and on the "interpretation" that different species give to these stimuli.

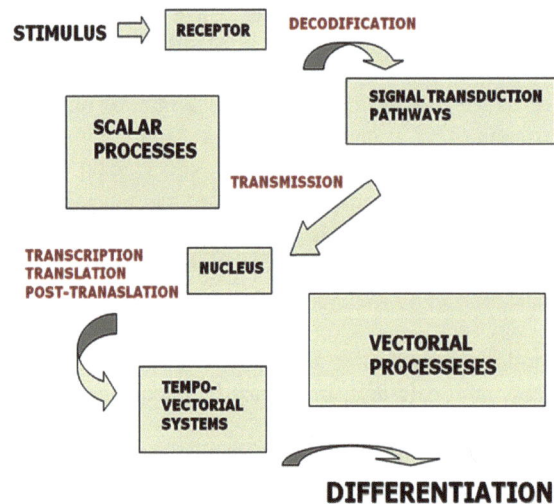

Figure 2: Schematic representation of the flow of information during differentiation. Stimuli react with cell receptors, unchaining a series of reaction by which the information is transferred by special transduction pathways to the nucleus, changing the running program of development through alterations in gene expression that lead to the establishment of a new developmental program. These changes include alterations in both scalar and vectorial processes.

Temperature

Both in *C. albicans*, *U. maydis* as in other model organisms, temperature plays an important role in dimorphism. This characteristic is especially useful for human pathogenic fungi, which at 25-28°C grow in the mycelial form, whereas at 37-40°C, they do it in the yeast-like form. The mechanism by which temperature regulates dimorphism is not understood, but possibly, it requires the presence of other effector molecules, the most obvious candidates appear to be heat shock proteins (HSPs). Kobayashi *et al.* mentioned that heat shock stimulated the HSP conserved system (HSPs 60-100 kDa) [24]. However, the authors considered that these proteins were not the effectors because HSPs synthesis can be induced in different organisms at different temperatures. So, they suggested that the specific receptor was located in the membrane and was affected by changes in its fluidity, exposing the active site. In this sense, it has been observed that the addition of oleic acid to *Histoplasma capsulatum* cultures, changes the HSP82 expression [24]. Maresca and Kobayashi [25] suggested that membrane phospholipids can be the sensing molecules. The authors observed that changes in the incubation temperature of *H. capsulatum* originated changes in fatty acid composition, while expression of *H. capsulatum* HSP82 was induced or reduced by the addition of saturated or unsaturated fatty acid, respectively. Coupling of mitochondrial ATPase activity was also affected by the addition of saturated fatty acid [26], and addition of fatty acids modified the yeast-to-mycelium transition time; while palmitic acid reduced it (from 7-8 days to 3 days) the addition of oleic acid increased it. Δ9-Fatty acid desaturase plays a specific role in the membrane fluidity adaptation to incubation temperature changes. It was shown to be bound to the endoplasmic reticulum membrane trough two transmembrane domains. It was observed that a shift in temperature from 25 to 37°C (or *vice versa*) diminished or increased, respectively the encoding (*Ole1*) gene transcription [27].

Effect of pH

In experiments devoted to the identification of probable gene products involved in the mechanism by which pH regulates morphogenesis in *C. albicans*, Fonzi´s group analyzed genes differentially expressed in response to pH changes [28]. They identified a gene (*PHR1*) encoding a glycoprotein bound to the plasma membrane through a GPI (glycosyl-phosphatidylinositol) moiety. In contrast to the wild type, the homozygous mutant in this gene was unable to form mycelium when grown at alkaline pH. Further work led to the identification of a similar gene (*PHR2*) encoding another GPI-protein expressed at acid pH, whose mutants were also affected in dimorphism at acid pH [29].

Perhaps more relevant to the mechanism by which pH regulates the differentiation, are the results obtained by analyzing the proteins secretion in different organisms, which operates through a mechanism of controlled proteolysis of a specific transcription factor. This control system was observed originally in *Aspergillus nidulans*, where it was described that regulation of secretion of enzymes by pH involved the proteolytic conversion of a transcription pre-factor (named *pacC or Rim101*) to its active form [30-32]. A similar mechanism was later on identified in *Yarrowia lipolytica* [33], but it was demonstrated that this mechanism of regulation was not involved in the yeast-to-mycelium transition of *Y. lipolytica* [34]. On the other hand *C. albicans* mutants in the pathway are affected in mycelial growth [35]. It was proposed that this pathway was specific of Ascomycota, but our identification of a homologue of PacC/Rim101 in the phytopathogen Basidiomycota *U. maydis*, showed that this concept was not correct [36], mutants affected in the pathway showed although the homologous pathway, identified the existence of homologous genes, although in *U. maydis* lacked two of the components existing in Ascomycota [37], same as we found to occur in all Basidiomycota analyzed. Interestingly, we observed by mutant isolation and phenotypic analyses, that as occurred with *Y. lipolytica*, this pathway was not involved in dimorphism.

Atmosphere of Growth

As already indicated, the presence or absence of oxygen, and the concentration of CO_2 under anaerobic conditions affect the dimorphic transition. The case of Mucorales is probably the most studied one. In the case of *M. bacilliformis* it is only necessary to eliminate oxygen from the atmosphere and substitute it by nitrogen or other inert gas to obtain the yeast-like morphology. On the other hand, *M. rouxii* requires high levels of CO_2 in anaerobic conditions, or anaerobiosis plus high concentrations of a fermentative carbon source to obtain the yeast form. *M. rouxii* contains an active mitochondrial respiratory chain whose inhibition by addition of cyanide never could be total, due to the presence of the SHAM (salicylhidroxamic acid)-sensitive alternative respiratory pathway [38]. However, under these conditions, the organism grew in the yeast-like form, and the same occurred with a mutant deficient in cytochrome) aa_3 [39].

In the case of *H. capsulatum*, it was described that the rather slow transition from yeast to mycelium required the addition of cysteine, whose role was suggested to be to lower the Redox potential of the cell. The sum of these results led us to suggest that mitochondrial respiration was necessary for mycelial growth;

and that the role of CO_2 was to induce a change in the internal pH of the cell, whereas the high concentrations of the fermentative hexose were related to an increase in the levels of cAMP.

Other effectors cited to play important roles in dimorphism, especially in the case of fungal pathogens of humans, are steroid derivatives. Accordingly, the work of several authors has shown that 17 β-estradiol inhibits the mycelium-to-yeast transition in *P. brasiliensis* [40], and the incidence of fungal infections caused by this pathogen is much lower in women than in men, having been suspected that a hormonal factor plays an important role in susceptibility or resistance.

Decoding of the Stimulus

Decoding of the stimulus in order to carry the information to the nucleus occurs by the activity of the so-called signal transduction pathways. It is interesting to notice that despite the diversity of signals that can trigger the dimorphic transition; the pathways that transfer the inducing signals to the nucleus are highly conserved in fungi. Two signaling pathways controlling morphogenesis (pseudo mycelial development) were originally identified in the budding yeast *S. cerevisiae* as a model: the mitogen-activated protein kinase (MAPK) and the cAMP/PKA transduction pathways [41] and further analyses have allowed the identification and characterization of homologues of its components in different fungal species: *Mucor circinelloides* [42], *Yarrowia lipolytica* [43, 44], *Paracoccidioides brasiliensis* [45], *Cryptococcus neoformans*, *Candida albicans*, *Aspergillus nidulans*, *Neurospora crassa*, and other plant fungal pathogens as *Ustilago maydis*, *Magnaporthe grisea*, *Cryphonectria parasitica*, *Colletotrichum* and *Fusarium* species, and *Erisyphe graminis* [41, 46].

Because of the amount of literature on this matter, description of the operation of these pathways in all the dimorphic models studied is out of the objectives of this introductory chapter, and as has been the rule followed up to here we will present only a few examples and the general aspects of the topic.

PKA Pathway

At this point it is important to recall that the this pathway involves a kinase dependent on cAMP (pKA) made of regulatory and catalytic subunit of the kinase whose role is to phosphorylate proteins involved in the adaptation of the cell to the new conditions. Another member of the pathway is adenylyl cyclase, responsible for the synthesis of the second messenger cAMP that activates the kinase, plus the membrane receptors (in the case of external stimuli, such as changes in the quality and quantity of nutrients, light, pH, temperature, *etc.*, see above) and a trimeric GTPase responsible for the initiation of the chain of transfer of the stimulus. When the receptor receives a signal, it stimulates the dissociation of the α-subunit (Gα), of the heterotrimeric G proteins from the heterodimer βγ Gα (and more rarely the heterodimer) stimulates or inhibits the adenylyl cyclase, initiating the transfer of the signal. Studies that confirm the functioning of this pathway in fungi, have been performed in a variety of species, *e.g. M. grisea* and *U. maydis* [see 47-50], by means of isolation of mutants in several of its members. Despite the high conservation of components of the pathway in different fungal species, there may be a differential control in different species of fungi; for example cAMP activates filamentous growth in *S. cerevisiae,* and positively regulates virulence in *C. neoformans*, whereas cAMP plays a negative role inhibiting filamentous growth in *U. maydis* [41]. Also, PKA (that was activated by cAMP), can have both positive or negative roles, by activating or inhibiting transcriptional activators or repressors such as Flo8 (a transcription factor essential for hyphal development and virulence in *C. albicans*) and Sfl1 (a negative regulator of hyphal development in *C. albicans*). These results suggests that similar to the case of *S. cerevisiae*, a combination of dual control by activation and repression of Flo8 and Sfl1 may contribute to the fine regulatory network in *C. albicans* morphogenesis responding to different environmental cues [41, 51].

MAPK Pathway

The other signal transduction pathway that has been shown to be involved in dimorphism is the MAPK pathway, a system made by three protein kinases operating in succession, normally kept close together through their association with a scaffold protein, plus the signal receptor and a G protein, either heterotrimeric or a small Ras-like. The signal unlashes a chain of successive phosphorylation reactions by the protein kinases denominated in order: Mapkkk, Mapkk, and Mapk that at the end is responsible for the phosphorylation of the proteins that, as occurs in the present case of the PKA pathway, are responsible for the differentiation phenomenon, in the presence case a dimorphic transition. The operation of both pathways may be synergic, their added activity being necessary for the dimorphic switch, or antagonic, each one involved in the acquisition of a different morphology. In *C. albicans*, mutation studies have revealed that the operation of both pathways is necessary for the correct germ tube formation [52-56]. Also in *S. cerevisiae* the pathways operate sinergically for the formation of pseudomycelium [57, 58], and in a similar way, the conjunction of these pathways is necessary for apresorium formation and tube germination in the biotrophic barley powdery mildew *Blumeria graminis* [59]. On the other hand, in *U. maydis* the pathways are antagonic; the pKA pathway is involved in yeast cell formation, whereas the function of a MAPK pathway is necessary for mycelial growth [47, 58, 60-62]. A similar phenomenon was observed with *Y. lipolytica*, where development of mycelium required a functional MAPK pathway [44], and the yeast-like growth did not occur when the catalytic subunit of the PKA was mutated [43]. When both pathways were non functional, the organism adopted a yeast-like growth, suggesting that this is the default morphology of the fungus; and interestingly, it was observed that in contrast to their antagonic behavior in dimorphism, both pathways are required for mating [44].

Change of Program and Final Reaction

Contrasting with the great advance that we have witnessed in our knowledge of stimuli reception and transduction of the signal, the mechanism leading to change in program has advanced slowly. The main problem in this subject is the difficulty in analyzing a great number of changes occurring simultaneously, and to interpret the information recovered. In these studies the use of libraries and microarrays might the best source for information.

Unfortunately the data obtained using these methods are few and yet, not conclusive to understand the changes in the cell programming in response to environmental stimuli. Considering that use of microarray or DNA chip techniques for *Candida albicans* research started recently, it is not strange that it has focused mainly in the understanding of pathogenesis, antifungal susceptibility, and diagnosis and only marginally on cell biology of this human pathogen [65]. Thus Bahn *et al.*, [64] showed that genes controlling metabolic specialization, cell wall structure, ergosterol/lipid biosynthesis, and stress responses were modulated by cAMP during hypha formation. Furthermore, several previously uncharacterized downstream targets that were up- or down-regulated by the cAMP signaling pathway also were identified, confirming previous insights into the role of the cAMP signaling pathway, not only in the morphogenic transitions of *C. albicans* but also in adaptation to stress and for survival during host infections [63] identified 514 pH-responsive genes. Of these, several involved in iron acquisition were upregulated at pH 8, suggesting that alkaline pH induces iron starvation. Analysis of *rim101* null mutants indicated that it does not govern transcriptional responses at acidic pH, but it does regulate a subset of transcriptional responses at alkaline pH, including the iron acquisition genes.

In *S. cerevisiae*, studies have been conducted to analyze the expression of genes in response to a variety of environmental changes [66-69]. The results obtained showed the relation between transcription factors and the genes they regulate. However, the analysis of results did not include a determination of the effect of these conditions in dimorphism.Regarding the final reaction, as we mentioned before, this aspect has the cell wall as the target, since it is the cell wall the structure that provides the cell its shape. For a long time, we have known that the important difference in the cell wall from the yeast-like form of a certain fungus, and the one making its hyphal wall is not the chemical composition, but the way it is deployed. Bartnicki-García and Lippman [70] demonstrated beyond any reasonable doubt that yeast growth occurs almost isotropic, with the wall being synthesized all over the surface of the cell; whereas the mycelium has a polarized growth that occurs only at the hyphal tip. To distinguish quantitatively yeast and mycelial forms Merson-Davies and Odds introduced the "morphology index" concept originally developed for *C. albicans* [71]. The morphology index (Mi) takes into account the following parameters:

s = Diameter at septal junction

h = Maximum diameter of cell (length)

w = Minimal diameter of cell

These parameters are associated by the equation:

$Mi = 2 + 1.78 \ (\log \ (hs/w^2)$

Where a value of Mi = 4, corresponds to hyphae, whereas a value of Mi = 1, corresponds to yeast.

This formula does not seem to apply to all organisms and it is likely that empirical data must be applied to other organisms.

We now know that for the synthesis of the fungal cell wall, its components and biosynthetic enzymes are conveyed to the cell surface packed into vesicles and nicrovesicles as originally shown many years ago by Girbardt [72, 73], and by Grove and Bracker [74]. According to these investigators, these vesicles and microvesicles accumulate at the hyphal apex in the form of a special organelle denominated Spitzenkörper by Brunswick [75], whose structure was resolved at the EM level by Girbardt [72, 73], and by Grove and Bracker [74]. From these organelle, the vesicles and microvesicles move to the cell surface. The way in which vesicles, microvesicles and the Spitzenkörper generate the hyphal shape was approached by different authors and methods, until Bartnicki-García *et al.*, [76, 77], utilized a computer model to demonstrate that the random movement of vesicles and microvesicles from the Spitzenkörper to the surface starting from a point that coincided with the position of the Spitzenkörper generated a novel shape that they denominated hyphoid because it coincided with the hyphal shape. That the movement of vesicles and microvesicles depends on the cytoskeleton, probably actin cables is supported by most authors.

In these pages our objective was to give a general presentation of the dimorphic phenomena in fungi, The reader will find specific aspects of the different fungal models selected for this volume in the following chapters. Briefly we can conclude that the final reaction of the dimorphic switch depends on the operation of a mechanism that transforms the information received, into a mechanism that converts an isodiametric type of growth into a polarized one, and vice versa, following the concepts delineated above. It is obvious that the shape of the cell is not the only change that occurs during the dimorphic transition. We know that changes in a high number of metabolic pathways occur during the morphogenetic reactions. In the last years we have witnessed great advances in their identification and role, and we are confident that in a few more years we will have a deep knowledge of the mechanisms controlling the whole phenomenon of fungal dimorphism.

ACKNOWLEDGEMENTS

The original work of the authors here reported has been partially supported by Consejo Nacional de Ciencia y Tecnología (CONACYT), México JRH is Emeritus investigator of the Sistema Nacional de Investigadores, México.

REFERENCES

[1]   Szaniszlo PJ (ed). Fungal Dimorphism: with Emphasis on Fungi Pathogenic for Humans. Plenum Publishing Corporation, New York 1985.

[2]   Karkowska-Kuleta J, Rapala-Kozik M, Kozik A. Fungi pathogenic to humans: molecular bases of virulence of *Candida albicans, Cryptococcus neoformans* and *Aspergillus fumigatus*. Acta Biochim Pol 2009; 56: 211-224.

[3]   Klein BS, Tebbets B. Dimorphism and virulence in fungi. Curr Opin Microbiol 2007; 10: 314-319.

[4]   Berkeley MJ. On a confervoid state of Mucor clavatus, Lk. Mag Zool Botany 1838, 2: 340-343.

[5]   Bail, T. 1857. Ueber Hefe. Flora (Jena) 40: 417-444.

[6]   Bartnicki-Garcia S. Cell wall chemistry, morphogenesis, and taxonomy of fungi. Annu Rev Microbiol 1968; 22: 87-108.

[7]   Pasteur, L. Études sur la Bière. Gauthier-Villars. Paris. 1876.

[8]   Brefeld O. *Mucor racemosus* und Hefe. Flora (Jena) 1873, 56: 386-399.

[9]   Wehmer C. Unabhängigkeit der Mucorineengärung von Säuerstoffabschluss und Kugelhefe. Ber Deut Botan Ges 1905; 23: 122-125.

[10]   Bartnicki-Garcia S, Nickerson WJ. Induction of yeast-like development in *Mucor* by carbon dioxide. J Bacteriol 1962; 84: 829-840.

[11]   Nickerson WJ, Bartnicki-García S. Specific and general aspects of the development of enzymes and metabolic pathways. Physiol Rev 1964; 44: 289-371.

[12]   Garrod D, Ashworth JM. Development of the cellular slime mold *Dictyostelium discoideum*. In: Ashworth JM, Smith JE, Eds. Microbial Differentiation. Cambridge University Press, Cambridge. 1973; pp 407-435.

[13]   Harold FM. To shape a cell: an inquiry into the causes of morphogenesis of microorganisms. Microbiol Rev 1990, 54: 381-431.

[14]   Wright BE. Concepts of differentiation. In: Smith JE, Berry DR, Eds. The Filamentous Fungi, Vol. 3, Develop Mycology. Edward Arnold, London. 1978, pp 2-7.

[15]   Ruiz-Herrera J. El dimorfismo de los hongos como modelo de diferenciación bioquímica. In: Martuscelli J, Palacios de la Lama R, Soberón Acevedo G, Eds. Caminos en la Caminos en la Biología Fundamental. Eds. J. Martuscelli, R. Palacios de la Lama & G. Soberón Acevedo. Universidad Autónoma de México. pp 285-303.

[16]   Herrmann H, Toole M. Specific and general aspects of development of enzymes and metabolic pathways. Physiol Rev.1964, 44: 289-358.

[17]   Mitchell P. Coupling of phosphorylation to electron and hydrogen transfer by a chemi-osmotic type of mechanism. Nature 1961, 191: 144-148.

[18]   Alvarez FJ, Konopka JB. Identification of an *N*-acetylglucosamine transporter that mediates hyphal induction in *Candida albicans*. Mol Biol Cell 2007, 18: 965-975.

[19]   Simonetti N, Strippoli V, Cassone A. Yeast-Mycelial Conversion Induced by N-acetyl-D-glucosamine in *Candida albicans*. Nature 1974, 250: 344-346.

[20]   Buffo J, Herman MA, Soll DR. A characterization of pH-regulated dimorphism in *Candida albicans*. Mycopathologia 1984, 85: 21-30.

[21]   Soll DR. Candida albicans. In Szaniszlo PJ, Ed. Fungal Dimorphism with Emphasis on Fungi Pathogenic for Humans. Plenum Press, New York. 1985, pp 167-195.

[22]   Martinez-Espinoza AD, Ruiz-Herrera J, Leon-Ramirez CG and Gold SE. MAP kinase and cAMP signaling pathways modulate the pH-induced yeast-to-mycelium dimorphic transition in the corn smut fungus *Ustilago maydis*. Curr Microbiol 2004, 49: 274-281.

[23]   Ruiz-Herrera J, Leon CG, Guevara-Olvera L, Carabez-Trejo A. Yeast-mycelial dimorphism of haploid and diploid strains of *Ustilago maydis*. Microbiology 1995, 141: 695-703.

[24]   Kobayashi GS, Medoff G, Maresca B, Sacco M, Kumar BV. Studies on phase transition in the dimorphic pathogen *Histoplasma capsulatum*. In: Szaniszlo PJ Ed. Fungal Dimorphism with Emphasis on Fungi Pathogenic for Humans. Plenum Press, New York. 1985, pp. 69-91.

[25]   Maresca B, Kobayashi G. Changes in membrane fluidity modulate heat shock gene expression and produced attenuated strains in the dimorphic fungus *Histoplasma capsulatum*. Arch Med Res 1993, 24: 247-249.

[26]   Carratù L, Franceschelli S, Pardini CL, Kobayashi GS, Horvath I, Vigh L, Maresca B. Membrane lipid perturbation modifies the set point of the temperature of heat shock response in yeast. Proc Natl Acad Sci USA. 1996, 9: 3870-3875.

[27]   Martin CE, Oh CS, Jiang Y. Regulation of long chain unsaturated fatty acid synthesis in yeast. Biochim Biophys Acta. 2007, 1771: 271-285.

[28]   Saporito-Irwin SM, Birse CE, Sypherd PS, Fonzi WA. *PHR1*, a PH-regulated gene of *Candida albicans*, is required for morphogenesis. Mol Cell Biol 1995, 15: 601-613.

[29]   Mühlschlegel FA, Fonzi WA. PHR2 of *Candida albicans* encodes a functional homolog of the pH-regulated gene PHR1 with an inverted pattern of pH-dependent expression. Mol Cell Biol 1997, 17: 5960-5967.

[30]   Negrete-Urtasun S, Denison SH, Arst HN. Characterization of the pH signal transduction pathway gene *palA* of *Aspergillus nidulans* and identification of possible homologs. J Bacteriol 1997, 179: 1832-1835.

[31]   Arst HN, Peñalva MA. PH regulation in *Aspergillus* and parallels with higher eukaryotic regulatory systems. Trends Genet 2003, 19: 224-231.

[32]   Peñalva MA, Tilburn J, Bignell E, Arst HN Jr. Ambient pH gene regulation in fungi: making connections. Trends Microbiol 2008, 16: 291-300.

[33]   Lambert M, Blanchin-Roland S, Lavedic FL, Leprinle A, Gaillardin C. Genetic analysis of regulatory mutants affecting synthesis of extracelllular proteinases in the yeast *Yarrowia lipolytica:* identification of a *RIM101/pacC* homolog. J Bacteriol 1997, 17: 3966-3976.

[34]   González-López CI, Ortiz-Castellanos L, Ruiz-Herrera J. The ambient pH response Rim pathway in *Yarrowia lipolytica*: Identification of *YlRIM9* and characterization of its role in dimorphism. Curr Microbiol 2006, 53: 8-12.

[35]   Davis D, Wilson RB, Mitchell AP. Rim101- dependent and -independent pathways govern pH responses in *Candida albicans*. Mol Cell Biol 2000, 20: 971-978.

[36]   Aréchiga-Carvajal ET, Ruiz-Herrera J. The Rim101/PacC homologue from the basidiomycete *Ustilago* maydis is functional in multiple pH-sensitive phenomena. Eukaryot Cell 2005, 4: 999-1008.

[37]   Cervantes-Chávez JA, Ortiz-Castellanos L, Tejeda-Sartorius M, Gold S, Ruiz-Herrera J. Functional analysis of the pH responsive pathway Pal/Rim in the phytopathogenic basidiomycete *Ustilago maydis*. Fungal Genet Biol 2010, 47: 446-457.

[38]   Cano-Canchola C, Escamilla E, Ruiz-Herrera J. Environmental control of the respiratory system in the dimorphic fungus *Mucor rouxii*. J Gen Microbiol 1988, 134: 2993-3000.

[39]   Salcedo-Hernández R, Ruiz-Herrera J. Isolation and characterization of a mycelial cytochrome aa3 deficient mutant and the role of mitochondria in dimorphism of *Mucor rouxii*. Exp Mycol 1993, 17: 142-154.

[40]   Restrepo A, Salazar ME, Cano LE, Stover EP, Feldman D, Stevens DA. Estrogens inhibit mycelium-to-yeast transformation in the fungus *Paracoccidioides brasiliensis*: implications for resistance of females to paracoccidioidomycosis. Infect Immun 1984, 46: 346-53.

[41]   Lengeler KB, Davidson RC, D'souza C, Harashima T, Shen WC, Wang P, Pan X, Waugh M, Heitman J. Signal transduction cascades regulating fungal development and virulence. Microbiol Mol Biol Rev. 2000, 64: 746-785.

[42]   Wolff AM, Appel KF, Petersen JB, Poulsen U, Arnau J. Identification and analysis of genes involved in the control of dimorphism in *Mucor circinelloides* (syn. racemosus). FEMS Yeast Res. 2002, 2: 203-213.

[43]   Cervantes-Chávez JA, Ruiz-Herrera J. STE11 disruption reveals the central role of a MAPK pathway in dimorphism and mating in *Yarrowia lipolytica*. FEMS Yeast Res. 2006, 6: 801-815.

[44]   Cervantes-Chávez JA, Kronberg F, Passeron S, Ruiz-Herrera J. Regulatory role of the PKA pathway in dimorphism and mating in *Yarrowia lipolytica*. Fungal Genet Biol. 2009, 46: 390-399.

[45]   Fernandes L, Araújo MA, Amaral A, Reis VC, Martins NF, Felipe MS. Cell signaling pathways in *Paracoccidioides brasiliensis*--inferred from comparisons with other fungi. Genet Mol Res. 2005, 4: 216-231.

[46]   Lee N, D'Souza CA, Kronstad JW. Of smuts, blasts, mildews, and blights: cAMP signaling in phytopathogenic fungi. Annu Rev Phytopathol 2003, 41: 399-427.

[47]   Gold SE, Duncan G Barrett KJ, Kronstad JW. cAMP regulates morphogenesis in the fungal pathogen *Ustilago maydis*. Genes Dev 1994, 8: 2805-2816.

[48]   Choi W, Dean RA. The adenylate cyclase gene *MAC1* of *Magnaporthe grisea* controls appressorium formation and other aspects of growth and development. Plant Cell 1997, 9: 1973-1983.

[49]   Liu S, Dean RA. G protein α subunit genes control growth, development, and pathogenicity of *Magnaporthe grisea*. Mol. Plant-Microbe Interact 1997, 10: 1075-1086.

[50]   Regenfelder E, Spellig T, Hartmann A, Lauenstein S, Bölker M, Kahmann R. G proteins in *Ustilago maydis*: transmission of multiple signals? EMBO J 1997, 16: 1934-1942.

[51]   Li Y, Su C, Mao X, Cao F, Chen J. Roles of *Candida albicans* Sfl1 in hyphal development. Eukaryot Cell. 2007, 6: 2112-2121.

[52]   Liu H, Kohler J, Fink GR: Suppression of hyphal formation in *Candida albicans* by mutation of a *STE12* homolog. Science 1994, 266: 1723-1726.

[53]   Malathi K, Ganesan K, Datta A. Identification of a putative transcription factor in *Candida albicans* that can complement the mating defect of *Saccharomyces cerevisiae* ste12 mutants. J Biol Chem. 1994, 269: 22945-22951.

[54]   Braun BR, Johnson AD. Control of filament formation in *Candida albicans* by the transcriptional repressor TUP1. Science. 1997, 277: 105-109.

[55]   Dhillon NK, Sharma S, Khuller GK. Signaling through protein kinases and transcriptional regulators in *Candida albicans*. Crit Rev Microbiol. 2003, 29: 259-275.

[56]   Whiteway M. Transcriptional control of cell type and morphogenesis in *Candida albicans*. Curr Opin Microbiol. 2000, 3: 582-588.

[57] Gancedo JM. Control of pseudohyphae formation in Saccharomyces cerevisiae. FEMS Microbiol Rev. 2001, 25: 107-123.

[58] Kronstad J, De Maria A, Funnell D, Laidlaw RD, Lee N, Moniz de Sa M, Ramesh M. Signaling *via* cAMP in fungi: interconnections with mitogen-activated protein kinase pathways. Arch Microbiol 1998, 170: 395-404.

[59] Kinane J, Oliver RP. Evidence that the appressorial development in barley powdery mildew is controlled by MAP kinase activity in conjunction with the cAMP pathway. Fungal Genet Biol. 2003, 39: 94-102.

[60] Gold SE, Brogdon SM, Mayorga ME, Kronstad JW. The *Ustilago maydis* regulatory subunit of a cAMP-dependent protein kinase is required for gall formation in maize. Plant Cell 1997, 9: 1585-1594.

[61] Kruger J, Loubradou G, Regenfelder E, Hartmann A, Kahmann R. Crosstalk between cAMP and pheromone signaling pathways in *Ustilago maydis*. Mol Gen Genet 1998, 260: 193-198.

[62] Kahmann R, Basse C, Feldbrugge M. Fungal-plant signalling in the *Ustilago maydis*-maize pathosystem. Curr Opin Microbiol 1999, 2: 647-650.

[63] Bensen ES, Martin SJ, Li M, Berman J, Davis DA. Transcriptional profiling in *Candida albicans* reveals new adaptive responses to extracellular pH and functions for Rim101p. Mol Microbiol 2004, 54: 1335-1351.

[64] Bahn YS, Molenda M, Staab JF, Lyman CA, Gordon LJ, Sundstrom P. Genome-wide transcriptional profiling of the cyclic AMP-dependent signaling pathway during morphogenic transitions of *Candida albicans*. Eukaryot Cell. 2007, 6: 2376-2390.

[65] Garaizar J, Brena S, Bikandi J, Rementeria A, Pontón J. Use of DNA microarray technology and gene expression profiles to investigate the pathogenesis, cell biology, antifungal susceptibility and diagnosis of *Candida albicans*. FEMS Yeast Res 2006, 6: 987-998.

[66] Gunji W, Kai T, Takahashi Y, Maki Y, Kurihara W, Utsugi T, Fujimori F, Murakami Y. Global analysis of the regulatory network structure of gene expression in *Saccharomyces cerevisiae*. DNA Res. 2004, 11: 163-177.

[67] Gasch AP, Spellman PT, Kao CM. *et al.* Genomic expression programs in the response of yeast cells to environmental changes. Mol Biol Cell 2000, 11: 4241-4257.

[68] Causton HC, Ren B, Koh SS *et al.*, Remodeling of yeast genome expression in response to environmental changes. Mol Biol Cell 2001, 12: 323-327.

[69] Capaldi AP, Kaplan T, Liu Y, Habib N, Regev A, Friedman N, O'Shea EK. Structure and function of a transcriptional network activated by the MAPK Hog1. Nat Genet. 2008, 40: 1300-1306

[70] Bartnicki-Garcia S, Lippman E. Fungal morphogenesis: cell wall construction in *Mucor rouxii*, Science. 1969, 165: 302-304.

[71] Merson-Davies LA, Odds FC. A Morphology Index for Characterization of Cell Shape in *Candida albicans*. J Gen Microbiol 1989, 135: 3143-3152.

[72] Girbardt M, Lebendbeobachtungen on *Polystictus versicolor* (L.). Flora 1955, 142: 540-563.

[73] Girbardt M. Der Spitzenkorper von *Polystictus versicolor* (L.). Planta 1957, 50: 47-59.

[74] Grove SN, Bracker CE. Protoplasmic organization of hyphal tips among fungi: vesicles and Spitzenkorper. J Bacteriol 1970, 104: 989-1009.

[75] Brunswick H. Untersuchungen uber Geschlechts und Kernverhaltnisse bei der Hymenomyzetengattung *Coprinus*, In Goebel K, Ed. Botanische Abhandlungen. Gustav Fisher, Jena, 1924, pp 5: 1-152.

[76] Bartnicki-Garcia S, Hergert F, Gierz G. A novel computer model for generating cell shape: application to fungal morphogenesis, In, Kuhn P, *et al.*, Eds. Biochemistry of Cell Walls and Membranes. Springer-Verlag, Heidelberg, 1989a, pp. 43-60.

[77] Bartnicki-Garcia S Hergert F, Gierz G. Computer simulation of fungal morphogenesis and the mathematical basis for hyphal (tip) growth. Protoplasma 1989b, 153: 46-57.

CHAPTER 2

Paracoccidioides, One Genus, More Than One Species: *P. brasiliensis* Cryptic Species and *P. lutzii.* Dimorphism, Morphogenesis, Phylogeny

Gioconda San-Blas[*] and Gustavo Niño-Vega

Laboratorio de Micología, Centro de Microbiología y Biología Celular, Instituto Venezolano de Investigaciones Científicas (IVIC), Apartado 20632,Caracas 1020A, Venezuela

Abstract: *Paracoccidioides brasiliensis*, a pathogenic dimorphic fungus geographically restricted to Latin America, is the causative agent of paracoccidioidomycosis, one of the most recurrent systemic mycoses in the region. Since its first description in 1908 by Adolpho Lutz in Brazil, *P. brasiliensis* has been considered the only species within the genus Paracoccidioides. Recent phylogenetic data have revealed the presence of cryptic species, namely *P. brasiliensis* S1, PS2 and PS3, and a well separated species, labeled *P. lutzii* as a tribute to Lutz. In this review, we bring together information that supports the existence of these new phylogenetic species, as well as summarize extensive works published on molecular aspects of dimorphism and morphogenesis done in classical strains of *P. brasiliensis* and also in *P. lutzii* (isolate Pb01). In doing so, we will attempt to analyze possible differences in their metabolic pathways, that may contribute to facilitate species differentiation and help researchers and clinicians to better understand the variety of pathologies so far reported in paracoccidioidomycosis.

Keywords: *Paracoccidioides brasiliensis, Paracoccidioides lutzii,* dimorphism, morphogenesis, molecular taxonomy, phylogenetic species, differentiation, paracoccidioidomycosis, adolpho lutz, mycelium, yeast, morphogenesis, conidia, cell wall.

INTRODUCTION

First described by Lutz [1], *Paracoccidioides brasiliensis* is the causative agent of paracoccidioidomycosis, a human systemic mycosis geographically confined to Latin America, where it constitutes one of the most prevalent deep mycoses [2]. The fungus is a pleomorphic organism that depends on the temperature of incubation for the expression of its morphology. *In vitro* at 37°C and in infected tissues, *P. brasiliensis* grows in the form of spherical or oval cells that vary in size from a few nanometers in diameter in young, recently separated buds to 10-30 μm or more in a mature yeast (Y) cell [3]. Yeast cells are multinucleate [4] and multiply by polar or multipolar budding. Buds are connected to the mother cell by narrow necks, giving the whole structure the characteristic appearance of a ship's wheel (Fig. **1A**), a shape considered the classical taxonomic and diagnostic characteristic of *P. brasiliensis*. The mycelial (M) form grows slowly at room temperature. Microscopic observations show septate, slender, and freely branching hyphae, 1-3 mm in width [4], with the appearance of interwoven threads (Fig. **1B**). Several types of conidia (intercalary, septate, pedunculate) are formed after 2 months of growth under conditions of nutritional deprivation [5, 6]. Conidia can germinate and produce Y cells at 37°C like those found in infected tissues, with the capacity to infect mice, or are able to produce germ tubes and branching mycelia if kept at 22°C.

Biochemical and Molecular Events in *P. brasiliensis* Morphogenesis and Dimorphism

Cell Wall structure

Earlier studies by Kanetsuna *et al.* [7] indicated that cell walls of *P. brasiliensis* varied in their composition according to the morphological phase. While chitin was common to both phases, glucose polymers were arranged mainly as α-1,3-glucan in the cell wall of the pathogenic Y phase (95% of glucans or 45% of the total cell wall) plus a small amount (5%) of β-1,3-glucan, whereas the latter was the only glucose polymer found in the M cell wall. Recent analysis of five *P. brasiliensis* strains, three from clinical cases (Pb73,

*Address correspondence to Gioconda San-Blas: Laboratorio de Micología, Centro de Microbiología y Biología Celular, Instituto Venezolano de Investigaciones Científicas (IVIC), Apartado 20632,Caracas 1020A, Venezuela. Email: sanblasg@ivic.gob.ve

José Ruiz-Herrera (Ed)

Pb381, Pb444), one from armadillo (Pb377) and one from soil (Pb300) [8] indicate that while their chemical composition followed somehow Kanetsuna's earlier data, figures varied slightly according to the isolate, *i.e.*, β-1,3-glucan amounts in the mycelial phase ranged from 20.2% (strain Pb444) to 31.4% (strain Pb73) of the total cell wall. Instead, the Y cell wall reduced this amount to 3.9-10.6% (isolates Pb377 and Pb73, respectively), while substituting this neutral polysaccharide by α-1,3-glucan (22.4 to 32.6% of the total Y cell wall; isolates Pb73 and Pb381, respectively). Chitin was around 2.5 times more abundant in the Y cell walls, as compared with their corresponding M cell walls. A recent revision of the chemical structure of α-1,3-glucan in the Y cell wall of *P. brasiliensis* (isolate Pb73) [9] confirmed the presence of α-1,3-glucan in the Y cell wall of *P. brasiliensis*, albeit with a different structural arrangement to that proposed before [10]. Our recent findings indicate a linear α-1,3-glucan, with occasional side chains of single α-1,4-linked glucose units. Some galactose and mannose (3-6%) are also found in *P. brasiliensis* cell wall. Galactomannan consists of a main chain of mannose linked through 1-6 linkages, to which galactofuranosyl residues are joined as non-reducing ends of the molecule [11]. This is an antigenic molecule that cross-reacts with other fungal mannans and galactomannans [12].

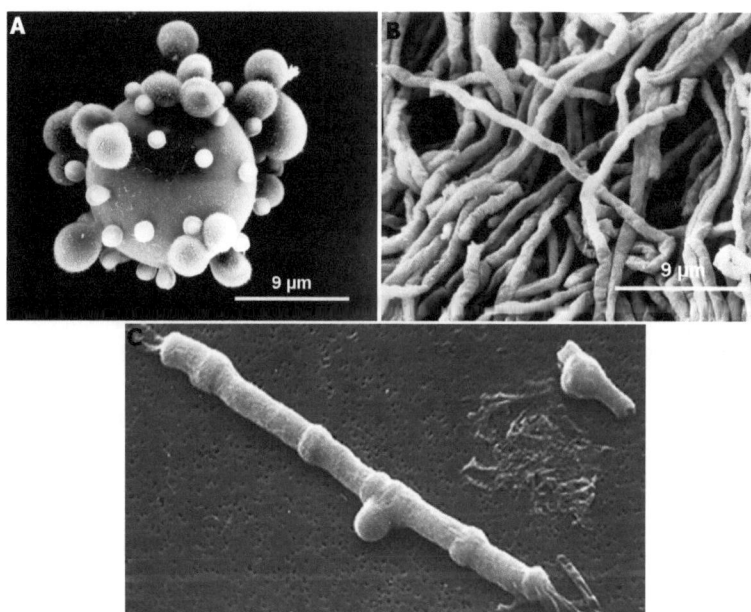

Figure 1: Electron micrographs of *P. brasiliensis* in the (A) yeastlike phase, (B) mycelial phase, (C) conidia. A and B, from our own collection; C, reproduced from [126] (x2000).

The mycelial cell wall has a single-layered cell wall, 0.08 to 0.15 mm thick, where chitin and β-glucan interwove [13]. The Y phase, instead, has a three-layered cell wall, 0.2 to 0.6 mm thick, in which two electron-dense and one electron-light zones alternate. The inner surface consists of a mixture of long fibers arranged in bundles (chitin) or isolated (β-glucan). The outer surface consisted of a mesh of short and thick fibrils of α-glucan, arranged in bundles, 20 to 240 μm wide and 200 μm long [4]. *P. brasiliensis* strains with a thicker layer of α-glucan are more virulent than those with thinner or inexistent layer of α-glucan in their cell wall. The peripheral α-glucan of *P. brasiliensis* may play a role as a protective layer against the host defense mechanisms [14], a kind of biochemical control of pathogenicity, also proposed years later for *Histoplasma capsulatum* [15, 16] and *Blastomyces dermatitidis* [17].

Cell Wall Synthesis

In *P. brasiliensis*, the synthesis *in vitro* of β-glucan by a membrane-bound β-1,3-glucan synthase requires UDP-glucose as the preferred nucleotide precursor [18, 19]. In common with some ascomycetes, the reaction is inhibited by GTP and other nucleotides [20], in sharp contrast to the general role played by these compounds as stimulators of fungal cell wall synthesis [21]. Only one gene (*PbFKS1*) can be traced in *P. brasiliensis* genome (Broad Institute, MIT and Harvard, Cambridge, MA;

www.broad.mit.edu/annotation/genome/paracoccidioides_brasiliensis.2). *PbFKS1* has an open reading frame of 5,942 bp; its complete sequence is interrupted by two putative introns [22]. The deduced sequence of 1,926 amino acids (predicted MW 212 kDa) shows 85% similarity to FksAp from *A. nidulans*, and 71% to Fks1p and Fks2p from *S. cerevisiae*. Unexpectedly, *PbFKS1* is more expressed when the culture grows in the yeast morphology or when the dimorphic M to Y transition reaches its final stages, instead of being expressed more actively in the β-1,3-glucan-related mycelial phase [9]. However, this result agrees with biochemical data previously reported by our own group [18], which indicates that the activity of β-1,3-glucan synthase *in vitro* is higher in particulate preparations from yeast cells than that of mycelial cultures.

The activity of glucan synthases depends on a previous activation by Rho GTPases. They function as on/off molecular switches, so that when bound to GTP, the GTPase is activated to affect downstream effectors, whereas when bound to GDP, the GTPase is deactivated [23]. Similarly to other fungal systems, it could be assumed that activation of β-1,3-glucan synthase in *P. brasiliensis* requires the participation of Rho1 as the GTPase regulatory subunit. *PbrRHO1*, in fact, is a gene whose putative product showed high identity with the regulatory subunit of several fungal β-1,3-glucan synthase complexes [9]. Its functionality was demonstrated by complementing a temperature-sensitive *S. cerevisiae rho1* null mutant with an expression vector containing the *P. brasiliensis RHO1* gene. Under non-inducible conditions, such transformants were able to grow at both 28 and 37°C, contrary to the control strain that only grew at 28°C, thus confirming that *P. brasiliensis RHO1* gene could rescue the temperature-dependent lethality due to the *rho1* mutation in *S. cerevisiae*.

P. brasiliensis α-1,3-glucan synthase gene (*AGS1*) presents six exons accounting for a putative coding region of 7293 bp, separated by five introns, all confirmed by RT-PCR [9]. The gene encodes a predicted protein of 2431 amino acids, with a calculated mass of 274.335 kDa. It shares a common structure with other fungal Ags (Mok) proteins [24-26] consisting of five domains: (i) a signal peptide (aa 1-20), (ii) an α-amylase homology domain (aa 66-521), (iii) a single transmembrane domain (aa 1073-1095), (iv) a glycosyl-transferase domain (aa 1178-1615), and (v) a carboxy-terminal domain (aa 1990-2422) with several membrane-spanning domains. The gene is expressed preferentially in the yeast phase, with only traces in the mycelial phase [9] and from 48 h into the mycelial to yeast transition, when most of the culture is already in the latter morphology (Fig. **1B**). *AGS1* is the only gene related to the synthesis of α-1,3-glucan, as in other pathogenic fungi such as *C. neoformans* and *H. capsulatum* [16, 27]. Instead, in *S. pombe* at least two different forms of cell wall α-1,3-glucan have been identified, one of them a linear α-1,3-glucan with occasional side chains of single α-1,4-linked glucose units in the sporulation process, very similar to the *P. brasiliensis* α-1,3-glucan [9]. The synthesis of *S. pombe* α-1,3-glucan is directed by Mok12 and Mok13; Mok12 aligns with *P. brasiliensis* Ags1 in a single cluster, suggesting a similar role in both fungi.

In *S. pombe*, another small GTPase (Rho2) is involved in the synthesis of cell wall α-1,3-glucan, *via* a Pck2 pathway [28, 29]. Comparison of the levels of expression of *P. brasiliensis AGS1* and *RHO2* in the M and Y stages of the fungus shows a direct correlation, suggesting a similar role of Rho2 in *P. brasiliensis*. Growth of *P. brasiliensis* Y phase cultures in YPG supplemented with 5% horse serum led to an increase in the α-1,3-glucan content of the Y cell wall, from 15 to almost 40% dry weight, accompanied by a slight increase in the level of expression of *AGS1* and a higher increase (near 40%) in the level of expression of *RHO2*. This suggests a post-transcriptional regulation of *P. brasiliensis AGS1*, as in *S. pombe*, where *AGS1* is post-transcriptionally regulated by Pck2 through the product of *RHO2* [30].

In all pathogenic fungi, chitin represents a major structural component of the cell wall with functions in fungal morphogenesis, wall integrity and conidiophore development [31]. Its synthesis is controlled by a multigene family, some of them redundant. Based on differences in regions of high sequence conservation, chitin synthases have been organized according to their amino acid sequences into two domains and seven classes [31, 32]. In *P. brasiliensis,* six chitin synthase genes have been accounted for; they represent different classes of enzyme, *i.e.*, *PbrCHS1* in class I, *PbrCHS2* in class II, *PbrCHS3* in class IV, *PbrCHS4* in class VII and *PbrCHS5* in class V [32-34] and a *CHS6* sequence whose translated product aligns with class VI chitin synthases from other fungi [35]; they help in the synthesis of chitin in amounts that comprise 43% of the dry weight of the wall of the pathogenic Y form and 13% of the M cell wall [36].

PbrCHS5 has a 5583 bp-long ORF, interrupted by three introns of 82, 87 and 97 bp. The deduced PbrCHS5 protein contains 1861 amino acids with a predicted molecular weight of 206.9 kDa. Its amino acid sequence is highly similar to other class V fungal chitin synthases [32, 34]. Two domains were identified, one towards the N-terminal end of the protein (aa 16 to 786), with partial identity to myosin motor-like domains, and a second one towards the C-terminal end (aa 1221 to 1752) with homology to fungal chitin synthases. Instead PbrChs4, while being a protein as large as PbrChs5, lacks sequences characteristic of myosin motors in its N-terminal region, a feature that classifies this chitin synthase apart from PbrChs5, in class VII [33, 34]. Sequence analysis of the 5'UTR resulted in overlapping with a previously reported sequence containing the *CHS4* gene [34], arranged in a head-to-head configuration with *CHS5*. So far, in *P. brasiliensis* only one single case of shared intergenic region had been reported between two genes, *MDJ1* and *LON* [37], whose products Mdj1 and Lon are a mitochondrial protein of the heat-shock protein family Hsp40 (Mdj1) and a conserved ATP-binding, heat-inducible serine proteinase (Lon), that colocalize with each other.

The nucleotide sequence of *PbrCHS2* contains a 2,540 bp ORF, with three introns, each of which has the characteristic splicing signal (lariat formation) observed in such sequences. The deduced amino acid sequence is 1043 residues long and predicts a 117 kDa protein that can be classified as a class II chitin synthase [38].

PbrCHS3 is the only one among the six chitin synthase genes reported so far in this species, to have a higher expression in the yeast phase and at the end of the mycelium-yeast transition [39]; all other *PbrCHSs* are preferentially expressed in the mycelial phase and the yeast-mycelium transition [32]; it contains a single open reading frame of 3817 bp with two introns (71 and 86 bp) and encodes a 1220 amino acid polypeptide with high similarity to other fungal chitin synthases. In *S. cerevisiae*, *CHS3* encodes the chitin synthase activity responsible for the synthesis of more than 90% of cellular chitin [40]. Expression of *P. brasiliensis CHS3* in a *S. cerevisiae chs3* null mutant demonstrates the functionality of the gene, inasmuch as an enhanced Calcofluor White staining, an increase in total chitin synthase activity and chitin content in its cell wall are observed [41].

A recent study [41], using the strain Pb01 (now the type strain for *Paracoccidioides lutzii*; see section 5) indicates that stressor agents such as Calcofluor White (CFW), Congo Red (CR), sodium dodecyl sulphate, KCl, NaCl or sorbitol trigger compensatory mechanisms against cell wall damage. 1,3-β-D-glucan synthase (*PbFKS1*), glucosamine-6-phosphate synthase (*PbGFA1)* and β-1,3-glucanosyltransferase (*PbGEL3)*as well as 1,3-β-D-glucan and *N*-acetylglucosamine (GlcNAc) residues in the cell wall were involved in such mechanisms. The relative expression level of *PbFKS1* increased after every treatment; *PbGFA1* was up-regulated after treatment with SDS and KCl and down-regulated by all other stressors. *PbGEL3* was up-regulated after treatment with CFW, NaCl and sorbitol. Together with such changes, modifications in the carbohydrate contents in the cell wall were recorded. β-1,3-D-glucan increased its contents after treatment with all stressor agents. An increase in GlcNAc contents after treatment with SDS or KCl suggested an increase in chitin. Such modifications were accompanied by a thickening in the yeast wall, as seen with confocal microscopy.

Other Metabolic Features in Morphogenesis

Cytoskeleton elements are tubulin and actin. In *P. brasiliensis*, the expression of α-tubulin increases during Y to M shift, only to decrease in the reverse process. Y cells expressed only the α_2 isoform of α-tubulin, whereas isoforms α_1 and α_2 were detected in mycelium [42]. The gene that encodes actin, *PbrACT-1*, is present as a single copy [43]. Its ORF encodes a putative protein of 375 amino acids, 41.4 kDa, and a >97% homology with other fungal actins. The N-terminal sequence corresponds to the γ-actin isotype. The gene is expressed preferentially in the yeast form of the fungus.

No significant differences were observed in the phospholipid species of both morphological phases of *P. brasiliensis* [44]. However, differences in the degree of unsaturation of fatty acids were recorded, so that the ratio of unsaturated to saturated fatty acids (UFA/SFA) was higher in the M form (2.10 in the M phase, and 0.49 in the Y phase) [44]; in this way, membranes are able to maintain normal basic permeability properties at the higher temperature in which Y cells are grown. The sterol biosynthetic pathway in the Y phase of *P. brasiliensis* goes one step forward in the classical route, ending up mostly in brassicasterol (ergosta-5,22-dien-3β-ol; 82% of the total sterols) instead of ergosterol (6%) [44, 45], a situation similar to that observed in vascular plants where the final products in sterol biosynthesis are Δ^5-24-alkyl sterol like stigmasterol, sitosterol and campesterol [46]. In the M phase, instead, ergosterol predominates (88 %), followed by low levels of brassicasterol (12 %).

Sphingolipids possibly modulate the activity of cell wall biosynthetic enzymes. In *P. brasiliensis*, glycosylinositol phosphorylceramides (GIPCs) consist almost exclusively of unbranched (t18:0) 4-hydroxysphinganine (phytosphingosine), in combination with saturated 2-hydroxy fatty acids, predominantly h24:0 [49], the mycelial cultures containing more total GPIs by weight than the yeast cells. The major neutral glycosylsphingolipids in both morphological phases are β-glucopyranosylceramides (GlcCer) with (4*E*,8*E*)-9-methyl-4,8-sphingadienine as a long chain base in combination with either *N*-2'-hydroxyoctadecanoate or *N*-2'-hydroxy-(*E*)-3'-octadecenoate, the latter an exclusive modification of fungal cerebrosides [48]. The mycelial GlcCer had both fatty acids in a 1:1 ratio, while the yeast GlcCer had only around 15% of the (*E*)-Δ^3-unsaturated fatty acid, a chemical dimorphism that was speculated to work as either a stabilizing factor in the maintenance of proper membrane structure or as a specific messenger in the Y to M conversion in a signalling cascade involving the activation or deactivation of the corresponding desaturase [48].

Sphingolipids and cholesterol, as important components of the cell membrane, may be organized in membrane rafts that play an essential role in different cellular functions, including host cell-pathogen interaction. In *P. brasiliensis*, the involvement of epithelial cell membrane rafts in the adhesion process of the pathogen and activation of cell signaling molecules was demonstrated after localization of ganglioside GM1, a membrane raft marker, at *P. brasiliensis*-epithelial cell contact sites, and by the inhibition of adhesion of this fungus to host cells pre-treated with cholesterol-extractor (methyl-β-cyclodextrin) or cholesterol-binding (nystatin) agents. At a very early stage of *P. brasiliensis*-cell interaction, the fungus promoted activation of Src-family kinases (SFKs) and extracellular signal-regulated kinase 1/2 (ERK1/2) of epithelial cells [49]. GM1 and GM3 may be involved in binding and/or infection by *P. brasiliensis*, as suggested by Ywazaki *et al.* [50], who found that anti-GM3 monoclonal antibody or cholera toxin subunit B (which binds specifically to GM1) reduced significantly fungal adhesion to fibroblast cells, by 35% and 33%, respectively. Such an effect is mainly due to the β-galactose or neuraminic acid at non-reducing end of glycosphingolipids. Other molecules, such as a 32-kDa haloacid dehalogenase protein, have been reported as putative adhesion molecules. The dehalogenase is a member of the superfamily of hydrolases, binds to extracellular matrix proteins and modulates the initial immune response for evasion of host defenses, therefore being involved in the virulence of this fungus [51].

The differential expression of a wide variety of proteins and genes has been reported. Silva *et al.* [52] reported the *hsp70* gene in *P. brasiliensis* which is differentially expressed during transition from the M to the Y form and after heat shock of yeast cells at 42° C. The gene encodes a 649 amino acid protein (predicted MW 70,461 kDa) of high identity with other members of the *hsp70* gene family, with six conserved sequence motifs characteristic of the Hsp70 family. Another heat shock gene, *hsp60*, is present as a single copy, and is composed of three exons and two introns [53]. The predicted HSP60 protein had two putative N-glycosylation sites, a result that is consistent with the finding of two HSP60 isoforms in *P. brasiliensis* yeast extracts, with differential N-glycosylation attributable to the presence of these two sites.

Kexin-like proteins are components of the subtilase family of proteinases, involved in the processing of proproteins to their active forms, inducing morphological cell defects. Such a gene was reported in *P. brasiliensis* [54]. *Pbrkex2* has an ORF of 2622 bp, interrupted by only one intron of 93 bp. It contains putative motifs for transcriptional factor binding sites, such as HSE-like motif that functions as a binding site for heat shock proteins with possible involvement in thermo-dependent processes.

Cell cycle and interaction between DNA replication, nuclei segregation and budding in *P. brasiliensis* have been poorly studied. Almeida *et al.* [55] focused on the characteristics of the cell cycle profile of *P.*

brasiliensis yeast cells during batch culturing and under the effects of benomyl, an antifungal drug known to promote a cell cycle arrest in the G_2/M phases of *Saccharomyces cerevisiae* and were able to demonstrate some level of independence between DNA duplication and cell division. Treatment with benomyl induced an arrest in the cell cycle profile and accumulation of cells in R_3 cells, a subpopulation that encloses a heterogeneous group of cells conformed by both single and multiple budding cells with 3 or more nuclei. These results suggested that even though benomyl progressively blocks nuclear division of *P. brasiliensis* yeast form, treated cells retained their capacity for DNA replication. *P. brasiliensis* yeast cells accumulated in stages of higher DNA content, rather than the G_2/M phases of the cell cycle, a fact consistent with their particular multinuclear and cellular division.

The glyoxylate cycle allows for the use of lipids in the synthesis of glucose *via* acetate generated in fatty acid β-oxidation and, additionally, seems to be involved in fungal pathogenicity [56]. Malate synthase is present in *P. brasiliensis*, where it plays a role in the primary regulation of carbon flux into the glyoxylate cycle, condensing acetyl-CoA from proline and purine metabolism to produce malate. With a calculated 539 amino acids and a molecular mass of 60 kDa, the gene that encodes it (*Pbmls*) has 1617 bp. The protein presents the MLSs family signature, the catalytic residues essential for enzymatic activity and the peroxisomal/glyoxysomal targeting signal PTS1 [57]. The enzyme could play a role in the binding of fungal cells to the host, thus contributing to the adhesion of fungus to host tissues and to the dissemination of infection [58].

Flavoprotein monooxygenases are oxidoreductases involved in a remarkably wide variety of oxidative reactions. One of them is the glycoprotein gp70, a concanavalin A-binding component recognized by about 96% of sera from untreated patients with PCM [59]. Its gene encodes for a protein sequence of 718 aminoacids; while its native form has a MW 79 kDa, it is reduced to about 70 kDa after endoglycosidase H-mediated N-deglycosylation. An increased PbGP70 transcript accumulation is observed under oxidative stress induced by H_2O_2, during fungal growth, and in macrophage phagocyted/bound yeasts. Therefore, a dual role in *P. brasiliensis* was suggested as (i) elicitor of an immune response in the host to antigenic and putative protective epitopes and (ii) as protector of the fungus against the oxidative stress generated either during growth or by the host immune responses, as in phagocytosis [59].

Catalases prevent the oxidative damage triggered by the reactive oxygen species of the host. Three catalases have been reported in *P. brasiliensis*, of which two [CatA and PbCatC] are monofunctional catalases and the third one [PbCatP], a catalase peroxidase [60]; additionally, *P. brasiliensis* has both cytosolic and peroxisomal catalase isoenzymes and a single cytochrome-c peroxidase [61]. PbCatA manifested higher activity in the mycelial phase, showed increased activity during M to Y transition and during conditions of endogenous oxidative stress. PbCatP showed higher activity in yeast cells since it is putatively involved in the control of exogenous reactive oxygen species. Like most monofunctional catalases, PbCatP is a homotetramer, resistant to inactivation by acidic conditions, temperature, and denaturants [62]. *P. brasiliensis* displays an oxidative stress response following phagocytosis by macrophages, inducing the expression of catalase A and P transcripts [61].

Enzymes whose encoding genes are differentially expressed through the morphogenetic process have been reported, *e.g.*, ornithine decarboxylase [ODC] [62], and calcineurin [64]. ODC is associated to the metabolism of polyamines [64]. Early work indicated that in *C. albicans*, *Mucor rouxii*, and *Y. lipolytica*, the activity of ODC was higher in the mycelial phase; in *P. brasiliensis*, instead, it is the yeast phase that shows a higher activity of the enzyme, either at the extreme morphologies or through mycelial to yeast transition [65]. However, *PbrODC* expression remained constant at all stages of the fungal growth, a result that suggests a post-transcriptional regulation of the *PbrODC* product [62]. Calcineurin, on the other hand, is a Ca^{2+}/calmodulin-dependent, serine/threonine-specific phosphatase essential for adaptation to environmental stress, growth, morphogenesis, and pathogenesis in many fungal species. Calcineurin controls hyphal and yeast morphology, M-Y dimorphism, growth, and Ca^{2+} homeostasis in *P. brasiliensis*, cyclosporine A being an effective chemical block for thermodimorphism in this organism, probably through regulation of plasma membrane Ca^{2+} channel Cch1p at the transcriptional level [62]. Carvalho *et al.* [66] found that several inhibitory drugs of the Ca^{2+}/calmodulin signalling pathways led to blockage of the

M to Y transition in *P. brasiliensis*, strongly suggesting an involvement of this pathway in morphogenesis. The calmodulin gene turned out to be 100% identical to that of *H. capsulatum* and was present as a single copy. Its deduced protein contained 149 amino acid residues with a molecular weight of 16.9 kDa and showed four Ca^{2+} binding sites per molecule.

Proteinases, namely serine proteinases, cysteine [thiol] proteinases, aspartic proteinases, and metalloproteinases, occur naturally in all organisms. A 66 kDa, N-glycosylated secreted aspartyl protease [PbSAP] of *P. brasiliensis* was identified and characterized [67]. The protein was located in the yeast cell wall and was also found in the supernatant. Venancio *et al.* [54] reported a kexin-like gene [*Pbkex2*] codifying for a kexin protein that belongs to the subtilase family of serine-proteinases. This proteinase family is involved in the processing of proproteins to their active forms. The nucleotide sequence of the *kex2* gene from *P. brasiliensis* [*Pbkex2*] contains an ORF of 2622 bp interrupted by one single 93 bp intron.

Also serine proteinases are the Lon proteins, with roles in the maintenance of mitochondrial DNA integrity and mitochondrial homeostasis. A *LON* gene homologue from *P. brasiliensis* [*PbLON*] was identified [68]. PbLon consists of 1,063 residues containing a mitochondrial import signal, a conserved ATP-binding site, and a serine catalytic motif. *PbLON* ORF is within a 3,369-bp fragment interrupted by two introns located in the 3' segment; an *MDJ1*-like gene was partially sequenced in the opposite direction, sharing with *PbLON* a common 5' untranslated region [37]. The authors propose that this chromosomal organization might be functionally relevant, since Mdj1p is a type I DnaJ molecule located in the yeast mitochondrial matrix and is essential for substrate degradation by Lon and other stress-inducible ATP-dependent proteinases.

An exocellular serine-thiol proteinase activity was reported by Matsuo *et al.* [69, 70] in the yeast phase of *P. brasiliensis* [PbST]. It was capable of cleaving proteins associated with the basal membrane, such as human laminin and fibronectin, type IV collagen and proteoglycans. Neutral polysaccharides but not sulfated glycosaminoglycans interacted with the proteinase, pointing to a novel modulation mechanism in *P. brasiliensis*, where a fungal polysaccharide-rich component can stabilize a serine-thiol proteolytic activity, which is possibly involved in fungal dissemination [70].

A particularly relevant *P. brasiliensis* glycoprotein is gp43. Gp43 is a secretory glycoprotein used as the main diagnostic and prognostic antigen so far characterized in *P. brasiliensis* because of its high levels of sensitivity and specificity for PCM patients' sera [71, 72]. Gp43 contains T cell epitopes that are protective to vaccinated mice [72], particularly the 15 aminoacid-long P-10 [73]. It also has adhesive properties to extracellular matrix proteins that may help fungal dissemination [74, 76]. The reference Pb339 strain has traditionally been employed in antigen preparation [76] because it secretes high amounts of gp43, as compared to other isolates [77]. The complete *PbGP43* ORF has originally been found in a cloned 3,800-bp *Eco*RI genomic region from the Pb339 isolate. It comprises 1,329 bp that contain a unique 78-bp intron [78]. In the phylogenetic studies performed by Matute *et al.* [79] (see section 5), *P. brasiliensis* Pb339 *PbGP43* was the most polymorphic gene in the multilocus analysis. Studies on sequence polymorphism in the *PbGP43* ORF [80] and its 5' intergenic proximal region [81], led to the definition of five genotypes [82]. The most polymorphic A genotype was detected in all 6 PS2 isolates [79]; it carried three substitutions in the 5' intergenic proximal region and up to 15 informative sites in the ORF, mostly concentrated in exon 2. Differences in gp43 expression could be related to differences in transcription regulation due to genetic polymorphisms in the *PbGP43* flanking regions [81]. Rocha *et al.* [83] characterized an extended 5' intergenic region up to 2,047 bp from Pb339 in comparison with other isolates and recognized some peculiar sequence organization. The authors cloned and sequenced the 5' intergenic region up to position -2,047 from *P. brasiliensis* Pb339 and observed that it is composed of three tandem repetitive regions of about 500 bp preceded upstream by 442 bp. Correspondent PCR fragments of about 2,000 bp were found in eight out of fourteen isolates; in PS2 samples they were 1,500-bp long due to the absence of one repetitive region, as detected in Pb3. In addition, they studied polymorphism in the 3' UTR and polyadenylation cleavage site of the *PbGP43* transcript. Accumulation of *PbGP43* transcripts was much higher in Pb339 than in Pb18 and Pb3, however they were similarly modulated with glucose. The amount of PbGP43 transcripts accumulated in *P. brasiliensis* Pb339 grown in defined medium was about 1,000-fold higher than in Pb18 and 120-fold higher than in Pb3 [83].

Studies on respiration, membrane potential, and oxidative phosphorylation in *P. brasiliensis* mitochondria revealed the presence of a complete (Complex I-V) functional respiratory chain and an alternative NADH-ubiquinone oxidoreductase [84]. Malate/NAD (+)-supported respiration suggested the presence of either a mitochondrial pyridine transporter or a glyoxylate pathway contributing to NADH and/or succinate production. Partial sensitivity of NADH/succinate-supported respiration to antimycin A and cyanide, as well as sensitivity to benzohydroxamic acids, suggested the presence of an alternative oxidase in the yeast form of the fungus. Further studies [85] indicated that M-to-Y transition was delayed by the inhibition of mitochondrial Complexes III and IV or alternative oxidase (AOX) and was blocked by the association of AOX with Complexes III or IV inhibitors. The expression of Pbaox was developmentally regulated through M-to-Y differentiation, wherein the highest levels were achieved in the first 24 h and during the yeast exponential growth phase; Pbaox was up regulated by oxidative stress.

Beyond Mycelium to Yeast Dimorphism: P. brasiliensis Conidia

Chlamidospores and arthrospores have been reported in *P. brasiliensis* [86, 87]. By far, the most extensively studied structure of this kind is the conidia, thanks to the efforts of Restrepo, McEwen *et al.*, in their Colombian laboratory. The first report on *P. brasiliensis* conidia [5] referred that growth of mycelia in poor culture media, namely water-agar, glucose-salts and yeast-extract, induced the formation of pear-shaped conidia 3.6 to 4.6 μm in length (Fig. **1C**). They share the ability of the parent mycelium to transform directly into multiple-budding yeast cells at 36 °C or to produce germ tubes and branching mycelia if kept at 22 °C [6]. In the process of conidia to yeast transition, some exclusive genes seem to be involved, as deduced from results obtained through an EST-Orestes library that was constructed and characterized [88]. As a result, 79 sequences were obtained, of which 39 (49.4%) had not been described previously in other libraries of this fungus. Two of these sequences are, among others, cholestanol delta-isomerase, and electron transfer flavoprotein-ubiquinoneoxidoreductase (ETF-QO). The other 40/79 (50.6%) sequences were shared with mycelia, yeast or mycelium to yeast transition libraries. In the process of conidia to mycelium transition, an EST library constructed to search for specific genes, produced eight sequences not previously described that may represent genes specific to the germination process [88] .

Transcriptome, Genome

Studies on the transcriptome of *P. brasiliensis* revealed expressed sequence tags (EST) clustered into 597 contigs and 1,563 singlets, making up a total of 2,160 genes, representing about one-quarter of the complete gene repertoire in *P. brasiliensis*. From this total, 1,040 were successfully annotated and 894 could be classified in 18 functional categories: cellular metabolism (44%); information storage and processing (25%); cellular processes-cell division, posttranslational modifications, among others (19%); and genes of unknown functions (12%) [89]. Almost simultaneously, Goldman *et al.* [90] obtained 13,490 ESTs corresponding to 4,692 expressed genes. Some of them (*e.g.* hydrophobin, isocitrate lyase, malate dehydrogenase, and alternative oxidase) displayed high mRNA expression in the mycelial phase, while others (*e.g.* ubiquitin, delta-9-desaturase, HSP70, HSP82 and HSP104) showed higher mRNA expression in the yeast phase. Molecular techniques such as microarrays and substraction hybridization allowed the identification of genes involved in basic and cell wall metabolism, sulfur metabolism, amino acid catabolism, signal transduction, growth and morphogenesis, protein synthesis, genome structure, oxidative stress response, and development genes that are preferentially expressed in the yeast phase [91-93]. Analyzing two cDNA populations of *P. brasiliensis*, one obtained from infected animals and the other an admixture of fungus and human blood that mimic the hematologic events of the fungal dissemination, Bailão *et al.* [94] were able to identify differentially expressed genes in host-fungus interaction. Genes related to iron acquisition, melanin synthesis and cell defense were specially upregulated in the mouse model of infection. The upregulated transcripts of yeast cells during incubation with human blood were those predominantly related to cell wall remodeling/synthesis. The expression pattern of genes was independently confirmed in host conditions, revealing their potential role in the infection process, and pointing to an eventual role in survival and growth strategies of *P. brasiliensis* in humans. The ability of *P. brasiliensis* in its yeast form to survive and replicate within the phagosome of nonactivated murine and human macrophages, is crucial to the development of disease, reason for which the fungus may have evolved mechanisms that counteract the constraints imposed by phagocytic cells. At least 152 genes primarily associated with glucose and amino acid limitation, cell wall construction, and oxidative stress, were differentially transcribed as a response to the environment of peritoneal murine macrophages [95]. Overexpressed transcripts from 4934 ESTs from *P. brasiliensis* yeast cells recovered from the livers of infected mice were related to glycolysis, amino acid biosynthesis, lipid and sterol metabolism, and suggested differential gene expression in response to the host milieu [96].

Contact with the NOK cells apparently induced alterations in the fungus, such as cellular extensions and cavitations, probably derived from modifications in the actin cytoskeleton [97]. Transcripts from 19 proteins involved in induction of cytokines, protein metabolism, alternative carbon metabolism, zinc transport proteins and stress response during contact with NOK cells were found.

One extraordinary step forward in the field of genomics has been the recent public release of the genome of three *P. brasiliensis* isolates: Pb18, Pb03 and Pb01, in an effort led by the Broad Institute, Massachusetts Institute of Technology MIT, Boston, that included all Latin American laboratories involved in molecular biological research of the fungus (Brazil, Colombia and Venezuela), under the *Paracoccidioides* Comparative Genome Analysis Project. Data can be found at http://www.broad.mit.edu/annotation/genome/ paracoccidioides_brasiliensis.2/MultiHome.html . Preliminary data-mining analyses indicated that the Pb01 strain does have important differences with the other two isolates, Pb18 and Pb03, particularly with regards to the genome size (32.94, 29.06 and 29.95 Mb, respectively) and number of genes (9132, 7875 and 8741 genes, respectively) [manuscript in preparation]. To this point we will return in section 5.

Electrophoretic karyotypes of 12 clinical and environmental *P. brasiliensis* isolates from different geographic areas indicated the possible existence of haploid and diploid [or aneuploid] isolates of the fungus [98]. Further studies by flow cytometry and comparison with previous electrophoretic data [55] revealed a genome size ranging from 26.3+/-0.1Mb (26.9+/-0.1fg) to 35.5+/-0.2Mb (36.3+/-0.2fg) per uninucleated yeast cell in 10 *P. brasiliensis* isolates. The analysis of intra-individual variability of the highly polymorphic *P. brasiliensis gp43* gene [80] indicated that only one allele was present; therefore, all isolates presented a haploid, or at least aneuploid, DNA content; no association was detected between genome size/ploidy and the clinical-epidemiological features of the isolates [55].

The Ascomycetous, Onygenalean *P. brasiliensis*

The distribution of two mating type loci (MAT1-1 or MAT1-2) in 71 *P. brasiliensis* isolates from various sources, and basal gene expression in some of them, suggests that sexual reproduction might occur in *P. brasiliensis* [99]. However, since no sexual structures have actually been detected, *P. brasiliensis* has been grouped within the Imperfect Fungi (Hyphomycetes or Mitosporic Fungi) by classical taxonomic criteria. First circumstantial evidence that it might be an Ascomycetous, came from Carbonell and Rodríguez [100], who reported simple septal pores and two-layered cell walls in *P. brasiliensis* hyphae, characteristic of ascomycetous fungi. However, the Y phase, multiplying by simultaneous multiple budding, had a mode of reproduction well known in zygomycetes but not in asco- or basidiomycetes, although the occasional rod-shaped invaginations visualized in the freeze-etching preparations of the plasma membrane of *P. brasiliensis* yeast cells suggested an asco or basidiomycetous origin, in contrast with the spherical invaginations found in zygomycetous plasma membranes [101].

Recent molecular evidence gives strong support to its classification as an Ascomycetous fungus, within the Onygenales [2]. Phylogeny of dermatophytes and dimorphic fungi based on the large subunit 28S ribosomal RNA sequence comparisons [102], the two ITS regions flanking the 5.8S rDNA as well as of the D1 and D2 domains of the LSU [103], and ODC [63], placed *P. brasiliensis* as belonging in the order Onygenales, family Onygenaceae (Phylum Ascomycota) together with *B. dermatitidis*, *H. capsulatum*, and *H. capsulatum* var. *duboisii* [102, 103] and in proximity with *Coccidioides immitis* [66]. Further SSU rDNA analyses coincided in locating *Lacazia loboi* and *P. brasiliensis* in the same taxonomic position [104]. Together, these data confirm *P. brasiliensis* as a member of the family Onygenaceae.

Other approaches complement the taxonomic information provided by molecular methods. With regards to *P. brasiliensis*, the water-soluble F1SS polysaccharides (mainly mannans and galactomannans obtained from the water-soluble fraction of the cell wall alkali extracts) which conforms a minor component of the cell wall (see section 1.1), have been recently proposed as a fungal taxonomic character [105], being useful in the definition of certain genera and determination of phylogenetic relationships in fungi. According to Leal *et al.* [105], in most members of the Onygenales, polysaccharides F1SS are composed of a chain of [1→6]-linked mannose with some moieties substituted by single residues or short chains of different sugars [106]. Recently, this polysaccharide has been fully characterized in the yeast and mycelial phases of *P. brasiliensis* [11]. The F1SS polysaccharide from the M form of *P. brasiliensis* has the characteristics assigned to F1SS of Onygenalean origin, that is, the distinguishing α-D-Gal*f*-[1→6]-α-D-Man*p*-[1→ side chains only found in polysaccharides isolated from species belonging to the family Onygenaceae, *e.g.*, *Ascocalvatia alveolata*, *Onygena equina* and *Aphanoascus terreus* [107], providing chemical evidence of *P. brasiliensis* as a member of the order Onygenales, phylum Ascomycota.

Paracoccidioides: More Than One Species Within the Genus

P. brasiliensis is confined to the Latin American region [2]. This fungus is considered clonal according to mycological criteria; at the same time, it shows extensive genetic variability when analyzed by molecular tools. RAPD analyses [108], RFLP [109], and partial sequences of some genes [80, 110] from a high number of *P. brasiliensis* isolates, revealed genetic variability and clusters correlated with geography or virulence [81, 111]. To characterize natural genetic variation and reproductive mode in this fungus, Matute *et al.* [79] analyzed *P. brasiliensis* phylogenetically in search of cryptic species and possible recombination using concordance and nondiscordance of gene genealogies with respect to phylogenies of eight regions in five nuclear loci. Their data indicate that this fungus consists of at least three distinct, previously unrecognized phylogenetic species (Fig. **2**): S1 (species 1 with 38 isolates from a broad geographical distribution within South America), PS2 (phylogenetic species 2 with five Brazilian and one Venezuelan isolate), and PS3 (phylogenetic species 3 composed solely of 21 Colombian isolates). S1 and PS2 were sympatric across their range, suggesting barriers to gene flow other than geographic isolation. All three species are capable of inducing disease in both humans and armadillos [81]. Matute *et al.* [112] also developed a marker system for DNA-based recognition of phylogenetic species S1 and PS2 in *P. brasiliensis*, based on microsatellites. Further search for evidence of positive selection in putative virulence factors within the *P. brasiliensis* species complex allowed Matute *et al.* [113] to report on the selection of 12 genes involved in different cellular processes, either antigenic or involved in pathogenesis. Two of them (*p27* and *gp43*) have unknown functions. All others were classified in four functional categories: metabolic related genes (*fas2*, *his1*), cell wall related genes (*fks*, *mnn5*, *ags1*), heat shock proteins, detoxification related genes (*tsa1*, *sod1*, *hsp88*) and signal transduction genes (*cdc42*, *cst20*).

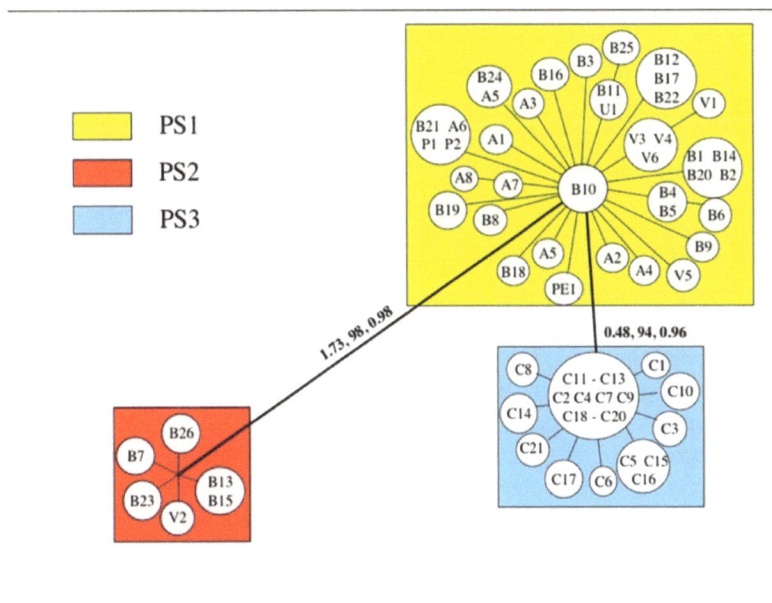

Figure 2: An un-rooted tree showing the partitions found in *P.brasiliensis*. When the eight loci are considered together the genealogies structure collapses into two branches, subdividing the population into three phylogenetic species. The branch lengths are drawn in scale only in the branches which connect the phylogenetic species [79]. Reproduced by permission.

Pbrchs2 was chosen by Matute *et al.* [114] to study background selection at the locus in *P. brasiliensis* species complex. For this, the DNA sequence for the *chs2* locus was determined in 67 samples. The sequenced region was 1162 bp in length, including 828 bp of exon and 334 bp of intron sequence. There were 21 sites polymorphic for nucleotide substitutions, 16 in exons and 5 in introns. No evidence for recombination within each one of the species for this gene was found.

Following the phylogenetic studies of Matute *et al.* [79], Carrero *et al.* [115] reported coding and non-coding regions from various genes and the ITS region in 21 isolates of *P. brasiliensis*, seven of them new. This study showed that the majority of the sequences used by Matute *et al.* [79] and those used in this study, grouped within two (S1 and PS3) of the three clades proposed by these investigators. However, one *P. brasiliensis* isolate, Pb01, was placed at the base of, and quite distant from, the three species reported by Matute *et al.* [79], clustering together with strain IFM 54648, an atypical strain isolated from a patient in the southern Brazilian region of Paraná [116]. The branch uniting the other *P. brasiliensis* isolates was well supported (>90%), whether the genes were highly conserved sequences, such as *HSP*, *Actin* or ribosomal genes, or more variable protein coding genes. This finding suggested the possibility of more than three phylogenetic species in *P. brasiliensis*. This may also suggests that Pb01-like isolates may have been genetically isolated from the other *P. brasiliensis* strains for a considerable period [115].

Further work [117] gave strength to this hypothesis, once the identification of 17 isolates, out of 88 samples, genotypically similar to Pb01, allowed their grouping as Pb01-like isolates (Fig. **3**). As a first approach to tell both sets apart, primers selected from the *hsp70* gene [52] were used. But definitive proof came from concordance and non-discordance analysis of multiple gene genealogies for phylogenetic species recognition (GCPSR) [117]. In this way, Pb01-like isolates are now considered a new phylogenetic species distinct from the S1, PS2 and PS3 clades previously reported by Matute *et al.* [79], since it is strongly supported by all independent and concatenated genealogies, with highly significant values of posterior probability (1.0) and bootstrap agreement (100%). The speciation event that defined this new phylogenetic group is sympatric relative to S1 and PS2. The two separate groups that include S1, PS2, PS3 on one side and Pb01-like on the other, were highly divergent [117]. Also, Pb01-like isolates present particular morphological characteristics when compared to the phylogenetic species S1, PS2, and PS3, namely, a more extended range area of yeast cell and distinctive elongated morphologies in its conidia. Additionally, data obtained from the Broad Institute project (see section 3) also point to genomic differences between isolate Pb01 and the other two isolates Pb03 (PS2) and Pb18 (S1) whose genomes have been decoded.

Based on molecular phylogenetic data, distinctive morphological characters and a long period of genetic isolation (>30 million years) that set the two groups apart, the Pb01-like clade may be considered a new phylogenetic species, and we have proposed the binomial name *Paracoccidioides lutzii* [117], whose specific descriptor is a tribute to Adolpho Lutz, the Brazilian researcher who first reported *P. brasiliensis* in 1908.

Additional information on transposable elements (TEs) validates the existence of several species within the genus *Paracoccidioides*. Marini *et al.* [118] reported 7 new families of DNA transposons and one new subfamily of Mariner-1_AF belonging to the *Tc1/mariner* superfamily. Remarkably, some Trem (Transposable element mariner) elements appear that may be active in the genomes of representative isolates of 2 different *Paracoccidioides* phylogenetic lineages. The finding of active autonomous DNA transposons in *P. brasiliensis* may have implications for an understanding of the evolutionary processes underlying the diversification of this group. Furthermore, transposons are efficient vectors for introducing foreign DNA into cells. Them elements represent a significant proportion of the genome-around 1% of the genome of *P. brasiliensis* (isolates Pb03 and Pb18) and 0.6% of *P. lutzii* (Pb01)-indicating their successful proliferation in the genome. The DNA transposons TremC and TremH were identified in all the isolates tested, indicating that these elements would have already been present in a hypothetical common ancestor rather than have been acquired horizontally after the 3 phylogenetic species diverged from one another. In contrast, TremE was only found in *P. lutzii*; TremA, TremB, and TremF were found in the S1 and PS2 isolates, while the element TremD was almost exclusively found in S1 isolates, suggesting that these DNA transposons could have been acquired horizontally after separation of the 3 phylogenetic species. Although TremD and TremE share approx 70% similarity at the amino acid level, they were found exclusively in the S1 and *P. lutzii* isolates, respectively, which are phylogenetically distant. This suggests that these elements would have been acquired by horizontal transfer after the split between S1/PS2 (*P. brasiliensis*) and *P. lutzii*.

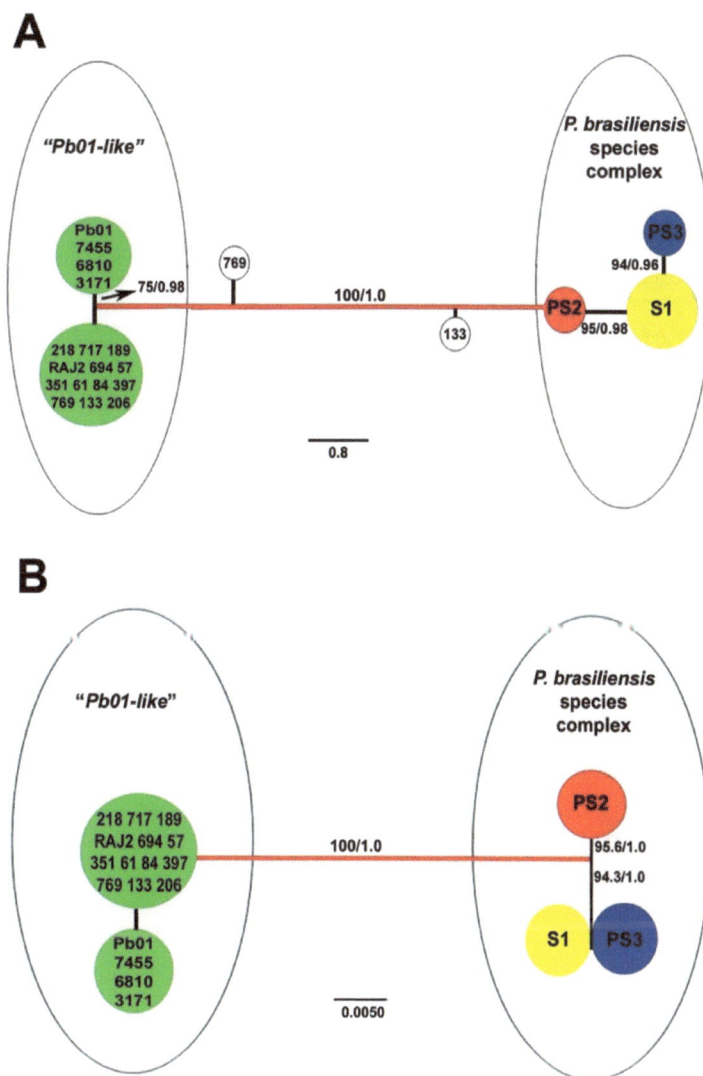

Figure 3: Bayesian unrooted phylograms showing the relationship between the isolates from the three *P. brasiliensis* phylogenetic species S1, PS2, PS3 and the isolates from the Pb01-like cluster, now *P. lutzii*. (A) Eight concatenated loci from dataset 1 (*fks*-exon2, *fks*-exon3, *chs*2-exon2-4, *gp43*-promoter-exon1, *gp43*-exon2, *arf* and alpha-tubulin. (B) Five concatenated loci from dataset 2 (hydrophobin-3'UTR, hydrophobin-5'UTR, *hsp70*-5'UTR and intron 1, *kex* intron 1 and ITS 1/2 + 5.8S) [117]. Reproduced by permission.

Do both species hold differences in their metabolic pathways? So far, studies on Paracoccidioides have been done mainly in *P. brasiliensis*. However, studies by Felipe's and Soares' groups have been carried out with isolate Pb01 (now *P. lutzii*) [119]. This means that a reevaluation of their results is in order, in an attempt to analyze possible differences in the expression of genes involved in a variety of metabolic pathways reflecting divergent characteristics in their evolution towards separate Paracoccidioides species.

Transcriptome analysis and molecular studies on a variety of metabolic pathways (glyoxalate, sulphur, carbon metabolism, and others) indicate that both species have, as expected, substantially similar metabolisms. However, differences can be spotted, in addition to those already mentioned. Fructose, galactose and glycerol are utilized as carbon sources by isolate Pb01 (*P. lutzii*) [89, 119] and not by isolate Pb18 (*P. brasiliensis*) [91]. Both species tend to be more aerobic in their mycelial phase than in their yeast form, albeit with some differences in the expression of certain genes. Genes encoding isocitrate dehydrogenase and succinyl-coenzyme A synthase are overexpressed in *P. lutzii*, while the genes encoding alcohol dehydrogenase I and isocitrate lyase were up-regulated in yeast cells, suggesting that the yeast metabolism favoured fermentation and consequent production of ethanol [119]; minor alterations in the expression of isocitrate dehydrogenase and succinyl-coenzyme A synthase but not isocitrate lyase, were detected in *P. brasiliensis* [92]. A 2.5-fold increase in the expression of malate synthase reinforced the idea that the glyoxylate pathway is more active in the yeast form of *P. brasiliensis* as compared to *P. lutzii* [92, 119]. Depending on the species, an increased expression of some genes that encode proteins involved in amino acid catabolism were reported, particularly aromatic and branched-chain amino acids that are also involved in a variety of metabolic functions such as synthesis of branched chain fatty acids, precursors of fusel alcohols, precursors of melanin [92, 119]. While the branched-chain aminotransferase, the first enzyme involved in the degradation of their respective amino acids, is preferentially expressed in yeast cells [119], an intense overexpression of the branched-chain α-keto acid dehydrogenase complex responsible for the second degradative step after 48 hours into mycelium to yeast transition [92]. Until more transcriptome research is done with additional isolates of both species, we will only be able to speculate if such discrepancies and/or other ones yet to be detected represent actual trends in the evolution towards speciation within the genus Paracoccidioides.

Do phylogenetic fungal species induce different morphological features in the cell or pathological effects in the patient? Not necessarily. So far, no differences whatsoever have been recorded in *Coccidioides immitis* and *C. posadasii*, the causative agents of coccidioidomycosis, two well established species differentiated only by means of phylogenetic criteria [120, 121]. Whether morphological differences can be observed in *P. brasiliensis* and *P. lutzii*, is something that is currently under research (Teixeira *et al.*, manuscript under preparation). As for pathological features, scattered data [122, 123] point to regionally-related variations in PCM pathologies (*i.e.*, lymphatic-abdominal) recorded in Goias, Brazil (a region close to Mato Grosso where almost all *P. lutzii* isolates have been recovered from PCM patients; [117], pathologies unnoticed in other Brazilian or Latin American regions. Furthermore, a recent report [124] indicates that while the standard gp43 antigen prepared from *P. brasiliensis* Pb339 [76] is highly reactive against sera from Southern Brazilian PCM patients, none to low levels of serological reactivity in PCM patients from the Central-Northern region of Brazil (Mato Grosso, Rondonia States) are recorded with this antigen; on the contrary, when gp43 is prepared from local isolates of the fungus, they are highly reactive with sera from these Brazilian states (92%) but become less reactive (41%) with sera from Southern Brazilian patients [124]. These results merit a study to relate them to the recently recognized existence of *P. lutzii* and the possibility of differential serological reactions within both Paracoccidioides species. It is tempting to think that the more frequent presence of *P. lutzii* as the prevalent causative agent of PCM in Central-Northern Brazil may be related to these observations [Teixeira *et al.*, manuscript under preparation]. Hahn *et al.* [125] observed that patients whose isolates clustered by RAPD analysis in clade II (now known to be *P. lutzii* isolates) had a better response to trimethoprim-sulfamethoxazole drug therapy than those patients whose isolates clustered in clade I (now known to be *P. brasiliensis* S1), thus becoming the only possible association so far reported between genotype and drug therapy response. The impact these data impose on medical practice is reason enough to make clinicians and medical staff aware of the recently acknowledged existence of different species within the genus Paracoccidioides and the negative consequences of using the inadequate exoantigen for serological tests. Advances in this area will have a positive influence on the diagnosis and therapeutics of a disease that seriously affects Latin American rural populations.

REFERENCES

[1]   Lutz A. 1908. Uma micose pseudo-coccídica localizada na boca e observada no Brasil: Contribução ao conhecimento das hiphoblastomicoses americanas. Brasil Med 1908; 22: 121-4.

[2]   San-Blas G, Niño-Vega G, Iturriaga T. *Paracoccidioides brasiliensis* and PCM: Molecular approaches to morphogenesis, diagnosis, epidemiology, taxonomy and genetics. Med Mycol 2002; 40: 225-42.

[3]   Carbonell LM. Ultrastructure of dimorphic transformation in *Paracoccidioides brasiliensis*. J Bacteriol 1969; 100: 1076-82.

[4]   Queiroz-Telles F. *Paracoccidioides brasiliensis* ultrastructural findings. In: Franco M, Lacaz C, Restrepo-Moreno A, Del Negro G, Eds. Paracoccidioidomycosis. Boca Raton, Florida, CRC Press, 1994; pp. 27-47.

[5]   Bustamante-Simon B, McEwen JG, Tabares AM, Arango M, Restrepo-Moreno A. Characteristics of the conidia produced by the mycelial form of *Paracoccidioides brasiliensis*. Sabouraudia 1985; 23: 407-14.

[6]   Restrepo A, Salazar ME, Cano LE, Patiño MM. A technique to collect and dislodge conidia produced by *Paracoccidioides brasiliensis* mycelial form. J Med Vet Mycol 1986; 24: 247-50.

[7]   Kanetsuna F, Carbonell LM, Moreno RE, Rodríguez J. Cell wall composition of the yeast and mycelial forms of *Paracoccidioides brasiliensis*. J Bacteriol 1969; 97: 1036-41.

[8]   Rodríguez-Brito S, Niño-Vega G, San-Blas G. Caspofungin affects growth of *Paracoccidioides brasiliensis* in both morphological phases. Antimicrob Agents Chemother 2010; 54: 5391-4.

[9]   Sorais F, Barreto L, Leal JA, Bernabe M, San-Blas G, Niño-Vega GA. Cell wall glucan synthases and GTPases in *Paracoccidioides brasiliensis*. Med Mycol 2010; 48: 35-47.

[10]   Kanetsuna F, Carbonell LM, Azuma I, Yamamura Y. Biochemical studies on the thermal dimorphism of *Paracoccidioides brasiliensis*. J Bacteriol 1972; 110: 208-18.

[11]   Ahrazem O, Prieto A, San-Blas G, Leal JA, Jiménez-Barero J, Bernabé M. Structural differences between the alkali-extracted water-soluble cell wall polysaccharides from mycelial and yeast phases of the pathogenic dimorphic fungus *Paracoccidioides brasiliensis*. Glycobiology 2003; 13: 743-7.

[12]   Azuma I, Kanetsuna F, Tanaka Y, Yamamura Y, Carbonell LM. Chemical and immunological properties of galactomannans obtained from *Histoplasma duboisii, Histoplasma capsulatum, Paracoccidioides brasiliensis*, and *Blastomyces dermatitidis*. Mycopathol Mycol App. 1974; 54; 111-25

[13]   Carbonell LM, Rodriguez J. Mycelial phase of *Paracoccidioides brasiliensis* and *Blastomyces dermatitidis*: an electron microscope study. J Bacteriol 1968; 96: 533-43.

[14]   San-Blas G, San-Blas F, Serrano LE. Host-parasite relationships in the yeastlike form of *Paracoccidioides brasiliensis* strain IVIC Pb9. Infect Immun 1977; 15: 343-6.

[15]   Klimpel KR, Goldman WE. Cell walls from avirulent variants of *Histoplasma capsulatum* lack alpha-(1,3)-glucan. Infect Immun 1988; 56: 2997-3000.

[16]   Rappleye CA, Engle JT, Goldman WE. RNA interference in *Histoplasma capsulatum* demonstrates a role for alpha-(1,3)-glucan in virulence. Mol Microbiol 2004; 53: 153-65.

[17]   Hogan L, Klein B. Altered expression of surface alpha-1,3-glucan in genetically related strains of *Blastomyces dermatitidis* that differ in virulence. Infect Immun 1994; 62: 3543-6.

[18]   San-Blas G. Biosynthesis of glucans by subcellular fractions of *Paracoccidioides brasiliensis*. Exp Mycol 1979; 3: 249-58.

[19]   Sorais-Landáez F, San-Blas G. Localization of β-glucan synthase in membranes of *Paracoccidioides brasiliensis*. J Med Vet Mycol 1993; 31: 421-6.

[20]   San-Blas G, San-Blas F. Effect of nucleotides on glucan synthesis in *Paracoccidioides brasiliensis*. Sabouraudia 1986; 24: 241-3.

[21]   Szaniszlo PJ, Kang MS, Cabib E. Stimulation of β-1,3-glucan synthase of various fungi by nucleoside triphosphates. A generalized regulatory mechanism for cell wall biosynthesis. J Bacteriol 1985; 161: 1188-94.

[22]   Pereira M, Felipe MSS, Brígido MM, Soares CMA, Azevedo MO. Molecular cloning and characterization of a glucan synthase from the human pathogenic fungus *Paracoccidioides brasiliensis*. Yeast 2000; 16: 451-62.

[23]   Boureux A, Vignal E, Faure S, Fort P. Evolution of the Rho family of Ras-like GTPases in eukaryotes. Mol Biol Evol 2007; 24: 203-16.

[24]   Katayama S, Hirata D, Arellano M, Pérez P, Toda T. Fission yeast alpha-glucan synthase Mok1 requires the actin cytoskeleton to localize the sites of growth and plays an essential role in cell morphogenesis downstream of protein kinase C function. J Cell Biol 1999; 144: 1173-86.

[25]   Damveld RA, vanKuyk PA, Arentshorst M, Klis FM, van den Hondel CA, Ram AF. Expression of agsA, one of five 1,3-alpha-D-glucan synthase-encoding genes in *Aspergillus niger*, is induced in response to cell wall stress. Fungal Genet Biol 2005; 42: 165-77.

[26]   García I, Tajadura V, Martín V, Toda T, Sánchez Y. Synthesis of alpha-glucans in fission yeast spores is carried out by three alpha-glucan synthase paralogues, Mok12p, Mok13p and Mok14p. Mol Microbiol 2006; 59: 836-53.

[27]   Reese AJ, Doering TL. Cell wall alpha-1,3-glucan is required to anchor the *Cryptococcus neoformans* capsule. Mol Microbiol 2003; 50: 1401-09.

[28]   Arellano M, Valdivieso MH, Calonge TM, *et al. Schizosaccharomyces pombe* protein kinase C homologues, pck1p and pck2p, are targets of rho1p and rho2p and differentially regulate cell integrity. J Cell Sci 1999; 112: 3569-78.

[29]   Calonge TM, Nakano K, Arellano M, *et al. Schizosaccharomyces pombe* rho2p GTPase regulates cell wall alpha-glucan biosynthesis through the protein kinase pck2p. Mol Biol Cell 2000; 11: 4393-401.

[30]   Hirata D, Nakano K, Fukui M, *et al.* Genes that cause aberrant cell morphology by overexpression in fission yeast: a role of a small GTP-binding protein Rho2 in cell morphogenesis. J Cell Sci 1998; 111: 149-59.

[31]   Ruiz-Herrera J, San-Blas G. Chitin synthesis as target for antifungal drugs. Curr Drug Target Infect Disorders 2003; 3: 77-91.

[32]   Niño-Vega GA, Munro CA, San-Blas G, Gooday GW, Gow NA. Differential expression of chitin synthase genes during temperature-induced dimorphic transitions in *Paracoccidioides brasiliensis*. Med Mycol 2000; 38: 31-9.

[33]   Niño-Vega GA, Carrero L, San-Blas G. Isolation of the CHS4 gene of *Paracoccidioides brasiliensis* and its accommodation in a new class of chitin synthases. Med Mycol 2004; 42: 51-7.

[34]   Niño-Vega GA, Sorais F, San-Blas G. Transcription levels of CHS5 and CHS4 genes in *Paracoccidioides brasiliensis* mycelial phase, respond to alterations in external osmolarity, oxidative stress and glucose concentration. Mycol Res 2009; 113: 1091-6.

[35]   Tomazett PK, Cruz AH, Bonfim SM, Soares CM, Pereira M. The cell wall of *Paracoccidioides brasiliensis*: insights from its transcriptome. Genetics Mol Res 2005; 4: 309-25.

[36]   San Blas G, San Blas F. Biochemistry of *Paracoccidioides brasiliensis* dimorphism. In: Franco M, Lacaz C, Restrepo-Moreno A, Del Negro G, Eds. Paracoccidioidomycosis. Boca Raton, Florida, CRC Press, 1994; pp. 49-66.

[37]   Batista WL, Barros TF, Goldman GH, Morais FV, Puccia R. Identification of transcription elements in the 5' intergenic region shared by LON and MDJ1 heat shock genes from the human pathogen *Paracoccidioides brasiliensis*. Evaluation of gene expression. Fungal Genet Biol 2007; 44: 347-56.

[38]   Niño-Vega GA, Buurman ET, Gooday GW, San-Blas G, Gow NAR. Molecular cloning and sequencing of a chitin synthase gene (CHS2) of *Paracoccidioides brasiliensis*. Yeast 1998; 14: 181-7.

[39]   Barreto L, Sorais F, Salazar V, San-Blas G, Niño-Vega GA. Expression of *Paracoccidioides brasiliensis CHS3* in a *Saccharomyces cerevisiae chs3* null mutant, demonstrates its functionality as a chitin synthase gene. Yeast 2010; 27: 293-300.

[40]   Choi WJ, Santos B, Duran A, Cabib E. Are yeast chitin synthases regulated at the transcriptional or the posttranslational level? Mol Cell Biol 1994; 14: 7685-94.

[41]   Tomazett PK, Castro Nda S, Lenzi HL, de Almeida Soares CM, Pereira M. Response of *Paracoccidioides brasiliensis* Pb01 to stressor agents and cell wall osmoregulators. Fungal Biol 2011; 115: 62-9.

[42]   Silva WP, Soraes RBA, Jesuino RSA, Izacc SMS, Felipe MSS, Soares CMA. Expression of alpha tubulin during the dimorphic transition of *Paracoccidioides brasiliensis*. Med Mycol 2001; 39: 457-62.

[43]   Niño-Vega G, Pérez-Silva C, San-Blas G. The actin gene in *Paracoccidioides brasiliensis*: organization, expression and phylogenetic analyses. Mycol Res 2007; 111: 363-9.

[44]   San-Blas G, Urbina J, Marchán E, Contreras LM, Sorais F, San-Blas F. Inhibition of *Paracoccidioides brasiliensis* by ajoene is associated with blockade of phosphatidylcholine biosynthesis. Microbiology 1997; 143: 1583-6.

[45]   Visbal G, Alvarez A, Moreno B, San-Blas G. S-Adenosyl-L-methionine inhibitors delta(24)-sterol methyltransferase and delta(24(28))-sterol methylreductase as possible agents against *Paracoccidioides brasiliensis*. Antimicrob Agents Chemother 2003; 47: 2966-70.

[46]   Nes WD. Sterol methyl transferase: enzymology and inhibition. Biochim Biophys Acta 2000; 1529: 63-88.

[47]   Levery SB, Toledo MS, Straus AH, Takahashi HK. Structure elucidation of sphingolipids from the mycopathogen *Paracoccidioides brasiliensis*: an immunodominant β-galactofuranose residue is carried by a novel glycosylinositol phosphoceramide antigen. Biochemistry 1998; 37: 8764-75.

[48]   Toledo MS, Levery SB, Straus AH, *et al.* Characterization of sphingolipids from mycopathogens: Factors correlating with expression of 2-hydroxy fatty acyl (E)-Δ^3-unsaturation in cerebrosides of *Paracoccidioides brasiliensis* and *Aspergillus fumigatus*. Biochemistry 1999; 38: 7294-306.

[49]   Maza PK, Straus AH, Toledo MS, Takahashi HK, Suzuki E. Interaction of epithelial cell membrane rafts with *Paracoccidioides brasiliensis* leads to fungal adhesion and Src-family kinase activation. Microbes Infect 2008; 10: 540-7.

[50]   Ywazaki CY, Maza PK, Suzuki E, Takahashi HK, Straus AH. Role of host glycosphingolipids on *Paracoccidioides brasiliensis* adhesion. Mycopathologia 2010; Nov 3. [Epub ahead of print].

[51]   Hernández O, Almeida AJ, Gonzalez A, Garcia AM, Tamayo D, Cano LE, Restrepo A, McEwen JG. A 32-kilodalton hydrolase plays an important role in *Paracoccidioides brasiliensis* adherence to host cells and influences pathogenicity. Infect Immun 2010; 78: 5280-6.

[52]   Silva SP, Borges-Walmsley MI, Pereira IS, Soares CMA, Walmsley AR, Felipe MSS Differential expression of an *hsp70* gene during transition from the mycelial to the infective yeast form of the human pathogenic fungus *Paracoccidioides brasiliensis*. Mol Microbiol 1999; 31: 1039-50.

[53]   Izacc SMS, Gómez FJ, Jesuino RSA, *et al.* 2001. Molecular cloning, characterization and expression of the heat shock protein 60 gene from the human pathogenic fungus *Paracoccidioides brasiliensis*. Med Mycol 39: 445-55.

[54]   Venancio EJ, Daher BS, Andrade RV, *et al.* The *kex2* gene from the dimorphic and human pathogenic fungus *Paracoccidioides brasiliensis*. Yeast 2002; 19: 1221-31.

[55]   Almeida AJ, Matute DR, Carmona JA, *et al.* Genome size and ploidy of *Paracoccidioides brasiliensis* reveals a haploid DNA content: flow cytometry and GP43 sequence analysis. Fungal Genet Biol 2007; 44: 25-31.

[56]   Dunn MF, Ramírez-Trujillo JA, Hernández-Lucas I. Major roles of isocitrate lyase and malate synthase in bacterial and fungal pathogenesis. Microbiology 2009; 155: 3166-75.

[57]   Zambuzzi-Carvalho PF, Cruz AH, Santos-Silva LK, *et al.* The malate synthase of *Paracoccidioides brasiliensis* Pb01 is required in the glyoxylate cycle and in the allantoin degradation pathway. Med Mycol 2009; 47: 734-44.

[58]   Neto BR, Silva JD, Mendes-Giannini MJ, *et al.* The malate synthase of *Paracoccidioides brasiliensis* is a linked surface protein that behaves as an anchorless adhesin. *BMC Microbiol* 2009; 9: 272.

[59]   Maricato JT, Batista WL, Kioshima ES, *et al.* The *Paracoccidioides brasiliensis* GP70 antigen is encoded by a putative member of the flavoproteins monooxygenase family. Fungal Genet Biol 2010; 47: 179-89.

[60]   Chagas RF, Bailão AM, Pereira M, *et al.* The catalases of *Paracoccidioides brasiliensis* are differentially regulated: protein activity and transcript analysis. Fungal Genet Biol 2008; 45: 1470-8.

[61]   Dantas AS, Andrade RV, de Carvalho MJ, Felipe MS, Campos EG. Oxidative stress response in *Paracoccidioides brasiliensis*: assessing catalase and cytochrome c peroxidase. Mycol Res 2008; 112: 747-56.

[62]   Niño-Vega GA, Sorais F, Calcagno AM, *et al.* Cloning and expression analysis of the ornithine decarboxylase gene (PbrODC) of the pathogenic fungus *Paracoccidioides brasiliensis*. Yeast 2004; 21: 211-8.

[63]   Campos CB, Di Benedette JP, Morais FV, Ovalle R, Nobrega MP. Evidence for the role of calcineurin in morphogenesis and calcium homeostasis during mycelium-to-yeast dimorphism of *Paracoccidioides brasiliensis*. Eukaryot Cell 2008; 7: 1856-64.

[64]   Reyna-López GE, Ruiz-Herrera J. Specificity of DNA methylation changes during fungal dimorphism and its relationship to polyamines. Curr Microbiol 2004; 48: 118-23.

[65]   San-Blas G, Sorais F, San-Blas F, Ruiz-Herrera J. Ornithine decarboxylase in *Paracoccidioides brasiliensis*. Arch Microbiol 1996; 165: 311-6.

[66]   de Carvalho MJ, Amorim Jesuino RS, Daher BS, Silva-Pereira I, de Freitas SM, Soares CM, Felipe MS. Functional and genetic characterization of calmodulin from the dimorphic and pathogenic fungus *Paracoccidioides brasiliensis*. Fungal Genet Biol 2003; 39: 204-10.

[67]   Tacco BA, Parente JA, Barbosa MS, *et al.* Characterization of a secreted aspartyl protease of the fungal pathogen *Paracoccidioides brasiliensis*. Med Mycol 2009; 47: 845-54.

[68]   Barros TF, Puccia R. Cloning and characterization of a LON gene homologue from the human pathogen *Paracoccidioides brasiliensis*. Yeast 2001; 18: 981-8.

[69]   Matsuo AL, Tersariol II, Kobata SI, *et al.* Modulation of the exocellular serine-thiol proteinase activity of *Paracoccidioides brasiliensis* by neutral polysaccharides. Microbes Infect 2006; 8: 84-91.

[70]   Matsuo AL, Carmona AK, Silva LS, *et al.* C-Npys (S-3-nitro-2-pyridinesulfenyl) and peptide derivatives can inhibit a serine-thiol proteinase activity from *Paracoccidioides brasiliensis*. Biochem Biophys Res Commun 2007; 355: 1000-5.

[71]   Puccia R, Schenkman S, Gorin PA, Travassos LR: Exocellular components of *Paracoccidioides brasiliensis*: identification of a specific antigen. Infect Immun 1986; 53: 199-206.

[72]   Travassos LR, Rodrigues EG, Iwai LK, Taborda CP: Attempts at a peptide vaccine against paracoccidioidomycosis, adjuvant to chemotherapy. Mycopathologia 2008; 165: 341-52.

[73]   Marques AF, da Silva MB, Juliano MA, Munhoz JE, Travassos LR, Taborda CP: Additive effect of P10 immunization and chemotherapy in anergic mice challenged intratracheally with virulent yeasts of *Paracoccidioides brasiliensis*. Microbes Infect 2008; 10: 1251-8.

[74]   Gesztesi JL, Puccia R, Travassos LR, *et al.* Monoclonal antibodies against the 43,000 Da glycoprotein from *Paracoccidioides brasiliensis* modulate laminin-mediated fungal adhesion to epithelial cells and pathogenesis. Hybridoma 1996; 15: 415-22.

[75]   Mendes-Giannini MJ, Andreotti PF, Vincenzi LR, *et al.* Binding of extracellular matrix proteins to *Paracoccidioides brasiliensis*. Microbes Infect 2006; 8: 1550-9.

[76]   Camargo Z, Unterkircher C, Campoy SP, Travassos LR: Production of *Paracoccidioides brasiliensis* exoantigens for immunodiffusion tests. J Clin Microbiol 1988; 26: 2147-51.

[77]   Moura-Campos MC, Gesztesi JL, Vincentini AP, Lopes JD, Camargo ZP: Expression and isoforms of gp43 in different strains of *Paracoccidioides brasiliensis*. J Med Vet Mycol 1995; 33: 223-7.

[78]   Cisalpino PS, Puccia R, Yamauchi LM, *et al.* Cloning, characterization, and epitope expression of the major diagnostic antigen of *Paracoccidioides brasiliensis*. J Biol Chem 1996; 271: 4553-60.

[79]   Matute DR, McEwen JG, Puccia R, *et al.* Cryptic speciation and recombination in the fungus *Paracoccidioides brasiliensis* as revealed by gene genealogies. Mol Biol Evol 2006; 23: 65-73.

[80]   Morais FV, Barros TF, Fukada MK, Cisalpino PS, Puccia R. Polymorphism in the gene coding for the immunodominant antigen gp43 from the pathogenic fungus *Paracoccidioides brasiliensis*. J Clin Microbiol 2000; 38: 3960-6.

[81]   Carvalho K C, Ganiko L, Batista WL, *et al.* Virulence of *Paracoccidioides brasiliensis* and gp43 expression in isolates bearing known *PbGP43* genotype. Microbes Infect 2005; 7: 55-65.

[82]   Puccia R, McEwen JG, Cisalpino PS. Diversity in *Paracoccidioides brasiliensis*. The PbGP43 gene as a genetic marker. Mycopathologia 2008; 165: 275-87.

[83]   Rocha AA, Morais FV, Puccia R. Polymorphism in the flanking regions of the PbGP43 gene from the human pathogen *Paracoccidioides brasiliensis*: search for protein binding sequences and poly(A) cleavage sites. BMC Microbiol 2009; 9: 277.

[84]   Martins VP, Soriani FM, Magnani T, *et al.* Mitochondrial function in the yeast form of the pathogenic fungus *Paracoccidioides brasiliensis*. J Bioenerg Biomembr 2008; 40: 297-305.

[85]   Martins VP, Dinamarco TM, Soriani FM, *et al.* Involvement of an alternative oxidase in oxidative stress and mycelium-to-yeast differentiation in *Paracoccidioides brasiliensis*. Eukaryot Cell 2011; 10: 237-48.

[86]   San-Blas F. Ultrastructure of spore formation in *Paracoccidioides brasiliensis*. J Med Vet Mycol 1986; 24: 203-10.

[87]   Franco M, Sano A, Kera K, Nishimura K, Takeo K, Miyaji M. Chlamydospore formation by *Paracoccidioides brasiliensis* mycelial form. Rev Inst Med Trop Sao Paulo 1989; 31: 151-7.

[88]   García AM, Hernández O, Aristizabal BH, *et al.* Gene expression analysis of *Paracoccidioides brasiliensis* transition from conidium to yeast cell. Med Mycol 2010; 48: 147-54.

[89]   Felipe MS, Andrade RV, Petrofeza SS, *et al.* Transcriptome characterization of the dimorphic and pathogenic fungus *Paracoccidioides brasiliensis* by EST analysis. Yeast 2003; 20: 263-71.

[90]   Goldman GH, dos Reis Marques E, Duarte Ribeiro DC, *et al.* Expressed sequence tag analysis of the human pathogen *Paracoccidioides brasiliensis* yeast phase: identification of putative homologues of *Candida albicans* virulence and pathogenicity genes. Eukaryot Cell 2003; 2: 34-48.

[91]   Marques ER, Ferreira ME, Drummond RD, *et al.* Identification of genes preferentially expressed in the pathogenic yeast phase of *Paracoccidioides brasiliensis*, using suppression subtraction hybridization and differential macroarray analysis. Mol Genet Genomics 2004; 271: 667-77.

[92]   Nunes LR, Costa de Oliveira R, Leite DB, *et al.* Transcriptome analysis of *Paracoccidioides brasiliensis* cells undergoing mycelium-to-yeast transition. Eukaryot Cell 2005; 4: 2115-28.

[93]   Bastos KP, Bailão AM, Borges CL, *et al.* The transcriptome analysis of early morphogenesis in *Paracoccidioides brasiliensis* mycelium reveals novel and induced genes potentially associated to the dimorphic process. BMC Microbiol 2007; 7: 29.

[94]   Bailão AM, Schrank A, Borges CL, *et al.* Differential gene expression by *Paracoccidioides brasiliensis* in host interaction conditions: representational difference analysis identifies candidate genes associated with fungal pathogenesis. Microbes Infect 2006; 8: 2686-97.

[95]   Tavares AH, Silva SS, Dantas A, *et al.* Early transcriptional response of *Paracoccidioides brasiliensis* upon internalization by murine macrophages. Microbes Infect 2007; 9: 583-90.

[96]   Costa M, Borges CL, Bailão AM, *et al.* Transcriptome profiling of *Paracoccidioides brasiliensis* yeast-phase cells recovered from infected mice brings new insights into fungal response upon host interaction. Microbiology 2007; 153: 4194-207.

[97] Silva RP, Matsumoto MT, Braz JD, *et al.* Differential gene expression analysis of *Paracoccidioides brasiliensis* during keratinocytes infection. J Med Microbiol 2010; Nov 11. [Epub ahead of print]

[98] Feitosa Ldos S, Cisalpino PS, dos Santos MR, *et al.* Chromosomal polymorphism, syntenic relationships, and ploidy in the pathogenic fungus *Paracoccidioides brasiliensis*. Fungal Genet Biol 2003; 39: 60-9.

[99] Torres I, García AM, Hernández O, *et al.* Presence and expression of the mating type locus in *Paracoccidioides brasiliensis* isolates. Fungal Genet Biol 2010 47: 373-80.

[100] Carbonell LM, Rodriguez J. Mycelial phase of *Paracoccidioides brasiliensis* and *Blastomyces dermatitidis*: an electron microscope study. J Bacteriol 1968; 96: 533-43.

[101] Takeo K, Sano A, Nishimura K, *et al.* Cytoplasmic and plasma membrane ultrastructure of *Paracoccidioides brasiliensis* yeast phase cells as revealed by freeze-etching. Mycol Res 1990; 94: 1118-22.

[102] Leclerc MC, Philippe H, Guého E. Phylogeny of dermatophytes and dimorphic fungi based on large subunit ribosomal RNA sequence comparisons. J Med Vet Mycol 1994; 32: 331-41.

[103] Bialek R, Ibricevic A, Fothergill A, Begerow D. Small subunit ribosomal DNA sequence shows *Paracoccidioides brasiliensis* closely related to *Blastomyces dermatitidis*. J Clin Microbiol 2000; 38: 3190-3.

[104] Herr RA, Tarcha EJ, Taborda PR, *et al.* Phylogenetic analysis of *Lacazia loboi* places this previously uncharacterized pathogen within the dimorphic Onygenales. J Clin Microbiol 2001; 39: 309-14.

[105] Leal JA, Prieto A, Ahrazem O, Pereyra T, Bernabé M. Cell wall polysaccharides: characters for fungal taxonomy and evolution. Rec Res Develop Microbiol 2001; 5: 235-48.

[106] Bernabé M, Ahrazem O, Prieto A, Leal JA. Evolution of polysaccharides F1SS and proposal of their utilisation as antigens for rapid detection of fungal contaminants. EJEAFChe 2002; 1(1). http://ejeafche.uvigo.es

[107] San-Blas G, Prieto A, Bernabé M, *et al.* α-galf 1→6-α-mannopyranoside side chains in *Paracoccidioides brasiliensis* cell wall are shared by members of the Onygenales, but not by galactomannans of other fungal genera. Med Mycol 2004; 43: 153-9.

[108] Calcagno AM, Niño-Vega G, San-Blas F, San-Blas G. Geographic discrimination of *Paracoccidioides brasiliensis* strains by randomly amplified polymorphic DNA analysis. J Clin Microbiol 1998; 36: 1733-6.

[109] Niño-Vega GA, Calcagno AM, San-Blas G, *et al.* RFLP analysis reveals marked geographical isolation between strains of *Paracoccidioides brasiliensis*. Med Mycol 2000; 38: 437-41.

[110] Hebeler-Barbosa F, Montenegro MR, Bagagli E. Virulence profiles of ten *Paracoccidioides brasiliensis* isolates obtained from armadillos (*Dasypus novemcinctus*). Med Mycol 2003; 41: 89-96.

[111] Molinari-Madlum EEWI, Felipe MSS, Soares CMA. Virulence of *Paracoccidioides brasiliensis* isolates can be correlated to groups defined by random amplified polymorphic DNA analysis. Med Mycol 1999; 37: 269-76.

[112] Matute DR, Sepulveda VE, Quesada LM, *et al.* Microsatellite analysis of three phylogenetic species of *Paracoccidioides brasiliensis*. J Clin Microbiol 2006; 44: 2153-7.

[113] Matute DR, Quesada-Ocampo LM, Rauscher JT, McEwen JG. Evidence for positive selection in putative virulence factors within the *Paracoccidioides brasiliensis* species complex. PLoS Negl Trop Dis 2008; 2: e296.

[114] Matute DR, Torres IP, Salgado-Salazar C, Restrepo A, McEwen JG. Background selection at the chitin synthase II (chs2) locus in *Paracoccidioides brasiliensis* species complex. Fungal Genet Biol 2007; 44: 357-67.

[115] Carrero LL, Niño-Vega G, Teixeira MM, *et al.* New *Paracoccidioides brasiliensis* isolate reveals unexpected genomic variability in this human pathogen. Fungal Genet Biol 2008; 45: 605-12.

[116] Takayama A, Itano EN, Sano A, Ono MA, Kamei K. An atypical *Paracoccidioides brasiliensis* clinical isolate based on multiple gene analysis. Med Mycol 2010; 48: 64-72.

[117] Teixeira MM, Theodoro RC, de Carvalho MJ, *et al.* Phylogenetic analysis reveals a high level of speciation in the *Paracoccidioides* genus. Mol Phylogenet Evol 2009; 52: 273-83.

[118] Marini MM, Zanforlin T, Santos PC, *et al.* Identification and characterization of Tc1/mariner-like DNA transposons in genomes of the pathogenic fungi of the *Paracoccidioides* species complex. BMC Genomics 2010; 11: 130.

[119] Felipe MSS, Andrade RV, Arraes FB, *et al.* Transcriptional profiles of the human pathogenic fungus *Paracoccidioides brasiliensis* in mycelium and yeast cells. J Biol Chem 2005; 280: 24706-14.

[120] Fisher MC, Koenig GL, White TJ, *et al.* Biogeographic range expansion into South America by *Coccidioides immitis* mirrors New World patterns of human migration. Proc Natl Acad Sci USA 2001; 98: 4558-62.

[121] Fisher MC, Koenig GL, White TJ, Taylor JW. Molecular and phenotypic description of *Coccidioides posadasii* sp. nov., previously recognized as the non-Californian population of *Coccidioides immitis*. Mycologia 2002; 94: 73-84.

[122] Barbosa W, Daher R, Oliveira AR. Forma linfático abdominal da blastomicose sul-americana. Rev Inst Med Trop São Paulo 1968; 10: 16-27.

[123] Andrade ALSS. Paracoccidioidomicose linfático-abdominal. Contribuição ao seu estudo. Rev Patol Trop 1983; 12: 165-256.

[124] Batista Jr J, de Camargo ZP, Fernandes GF, *et al.* Is the geographical origin of a *Paracoccidioides brasiliensis* isolate important for antigen production for regional diagnosis of paracoccidioidomycosis? Mycoses 2010; 53: 176-80.

[125] Hahn RC, Macedo AM, Fontes CJF, *et al.* 2003. Randomly amplified polymorphic DNA as a valuable tool for epidemiological studies of *Paracoccidioides brasiliensis* J Clin Microbiol 41: 2849-54.

[126] Restrepo A, Aristizábal BH, González A *et al.* Características de las conidias de *Paracocidioides brasiliensis*. Vitae Academia Biomédica Digital 2000; 40 http://vitae.ucv.ve

CHAPTER 3

Dimorphism and Pathogenicity of the Opportunistic Ascomycota *Candida albicans*

Margarita Rodríguez-Kessler[1,*], María de la Luz Guerrero-González[2] and Juan Francisco Jiménez-Bremont[2]

[1]Facultad de Ciencias, Universidad Autónoma de San Luis Potosí (UASLP). San Luis Potosí, México and [2]División de Biología Molecular, Instituto Potosino de Investigación Científica y Tecnológica (IPICYT), San Luis Potosí, Mexico

Abstract: *Candida albicans* is an opportunistic human fungal pathogen that belongs to the *Saccharomycetaceae* family of ascomycota fungi. *C. albicans* is responsible for local and systemic infections, mainly in immunocompromised patients. Inside the host, it has the ability to form biofilms, to adapt to different environmental pressures and to switch between yeast and filamentous forms. The molecular mechanisms behind dimorphic transition in *C. albicans* and its relation to pathogenesis are scientific highlights, and many efforts have been done to understand and identify key regulators in this process. In the present chapter, we review many important regulators of yeast-to-hypha transition in *C. albicans*, including transcription factors, signaling mediated by cAMP and MAPK pathways, pH dependent morphological transitions and the role of important growth regulators such as polyamines. Even so information on new molecules is still needed to fully understand the mechanisms governing dimorphism in *C. albicans*.

Keywords: *Candida albicans,* dimorphism, morphogenesis, mycelium yeast, differentiation, morphogenesis, pathogenicity, MAP kinase, cAMP, PKA, pH effect.

INTRODUCTORY ASPECTS

Candida albicans, is an opportunistic, ascomycota fungus, usually found as a commensal of mucosal surfaces in the oral and vaginal cavities, and as part of the normal gastrointestinal flora in healthy humans [1]. Clinical importance of *C. albicans* focus on immunocompromised patients (due to tissue or organ transplantation, chemotherapy, acquired immunodeficiency syndrome, *etc*), which are particularly susceptible to infection by this fungus. In fact, *C. albicans* is considered the most common human fungal pathogen, and it has been documented as the fourth most frequent hospital acquired infection [1, 2]. Accordingly, *C. albicans* ability to form biofilms on abiotic surfaces (implants), to resist antifungal drugs (azoles), to switch between yeast and filamentous (pseudohyphae and true hyphae) forms, and to adapt to different environmental pressures inside the host, are key factors in their virulent behavior responsible for their local and systemic infections that account for substantial morbidity and mortality worldwide [1-6]. In this regard, *ca.* 30 to 50% mortality rate has been associated to systemic infections (candidiasis) [1, 2, 7]. One of the most important virulence traits of *C. albicans* is its capacity to switch between yeast and mycelial morphologies, a process that is reversible. Up to now, the molecular and regulatory mechanisms that control *C. albicans* morphological switches and that modulate its lifestyle as a successful commensal or pathogen are still not fully understood.

C. albicans is a diploid organism that belongs to the Saccharomycetaceae family and is closely related to other *Candida* species (*C. dubliniensis, C. tropicalis* and *C. parapsilosis*) that have undergone whole genome duplication [8]. Large scale sequencing projects have focused on the *C. albicans* strains SC5314 and WO-1. The first one, is a widespread strain used in molecular analyses and virulence tests in animal models [9], and the second one has the ability to switch between white and opaque morphologies (see

*Address correspondece to Margarita Rodriguez-Kessler: Facultad de Ciencias, Universidad Autónoma de San Luis Potosí. San Luis Potosí, México. Av. Salvador Nava s/n, Zona Universitaria C.P. 78290, San Luis Potosí, SLP, México; E-mail: mrodriguez@fc.uaslp.mx

section below), a phenotypic change associated with host specificity and mating [10]. Their complete genomic sequence is available at http://www.candidagenome.org/ and at http://www.broadinstitute.org/annotation/genome/candida_group/MultiHome.html, respectively. They consist of approximately 13.3-13.4 Mb encoding *ca.* 6,107 - 6,159 genes organized into 8 chromosomes [9-11]. In addition, *C. albicans* belongs to the CUG clade that has an abnormality in its genetic code in which the leucine codon CUG is read as serine [10, 12]. One of the main limitations in molecular genetic studies and functional analysis in *C. albicans* was the generation of homozygous mutants, following two successive rounds of transformation to delete both alleles in the diploid genome [13, 14]. In this sense, several molecular techniques were developed for gene manipulation in this organism and also to improve transformation protocols [15]. Important to mention are the development of auxotrophic markers (*e.g. URA3*), of dominant selectable markers conferring resistance to antibiotics (*e.g.*, *SAT1* gene conferring resistance to nourseothricin) and markers of conditional expression (*e.g.,* tetracycline-repressible or -inducible promoters) [16-18]. In addition, a fusion PCR method that generates a gene disruption fragment with stretches of homology (*ca.* 350-500 bp) that flank the target gene was also developed [19]. This method uses *HIS1, ARG4* and *LEU2* genes as markers for gene disruption, and has advantages regarding the usage of *URA3* gene, since deletion of the corresponding amino acids have minimal effects on virulence and are suitable for *in vivo* studies [reviewed in 1, 15]. In addition, *C. albicans* insertional mutants using the *UAU1*-marked Tn7 transposon, enabled the generation of libraries of homozygous mutants for the identification and functional characterization of genes involved in mycelial development, pH response, and biofilm formation, among others [13, 20-22]. Recently, tools for *in vivo* protein tagging using PCR-mediated homologous recombination of TAP, HA and MYC tags were developed and tandem affinity purification and chromatin immunoprecipitation analysis were achieved [23]. Furthermore, a β-galactosidase reporter system for gene expression analysis is also available [23, 24]. All these tools are of substantial assistance in the understanding of *C. albicans* biology.

In this chapter, special attention will be given to the molecular mechanisms involved in the yeast-to-hyphae transition process, and its importance in virulence in *C. albicans*.

Dimorphism in Fungi

Several fungal species exhibit a morphological transition (dimorphism), expressed as the capacity to grow in two distinct morphological forms: *e.g.*, as yeast cells or as filamentous hyphae [25]. This phenomenon is believed to constitute a mechanism of response to environmental conditions and represents an important attribute for the development of virulence by a number of human and plant pathogenic fungi, including *C. albicans, Histoplasma capsulatum, Blastomyces dermatitidis, Paracoccidioides brasiliensis, Coccidioides immitis, C. posadasii, Sporothrix schenckii, Penicillium marneffei, Ustilago maydis, Ceratocystis ulmi, Mycosphaerella graminicola*, among others [26, 27]. Several conditions have been described that induce the dimorphic transition of yeast-to-mycelium or mycelium-to-yeast, and switch between a saprophytic and a virulent phase of growth. Among these, changes in ambient pH, temperature, nutritional status (starvation), the gaseous atmosphere of growth (CO_2 content), or the presence of specific compounds in the culture media such as *N*-acetyl-D-glucosamine, L-proline, blood serum, lipids, *etc.* [25, 28, 29]. Quorum sensing systems, such as the production of farnesol in *C. albicans*, have also being described as regulators of dimorphic transition [30].

Dimorphism in Candida albicans

C. albicans is a commensal fungus of numerous warm-blooded animals, including humans; therefore it is mainly isolated from clinical samples rather than from environmental sources as occurs for other fungi. In most cases, *C. albicans* grows in the form of yeast, pseudohyphae or true hyphae. The ability to switch among these morphologies is often considered to be required for virulence although all forms have been identified during infection [31, 32]. Hyphae and pseudohyphae are invasive structures, and it has been documented that hyphal growth is necessary to evade phagocytes, to escape from blood vessels, and to colonize organs and abiotic surfaces forming biofilms [31, 32]. On the other hand, the yeast form may be involved in dissemination in the blood stream. Yeast cells reproduce by budding, while pseudohyphae result from daughter buds that elongate and form a division septum during mitosis but fail in separating from the mother cell, Fig. **1**. True hyphae are linear filaments without visible constrictions at the septa and result from germ tube growth. Several environmental conditions are known to promote the transition among the different morphological structures in *C. albicans*: yeast (growth below 30 °C, pH 4.0, quorum sensing), pseudohyphae (35 °C, pH 6.0, high phosphate) and hyphae (37 °C, pH 7.0, high CO_2, serum, *N*-acetyl-D-glucosamine, starvation and adherence) [29, 30, 32; Fig. **1**].

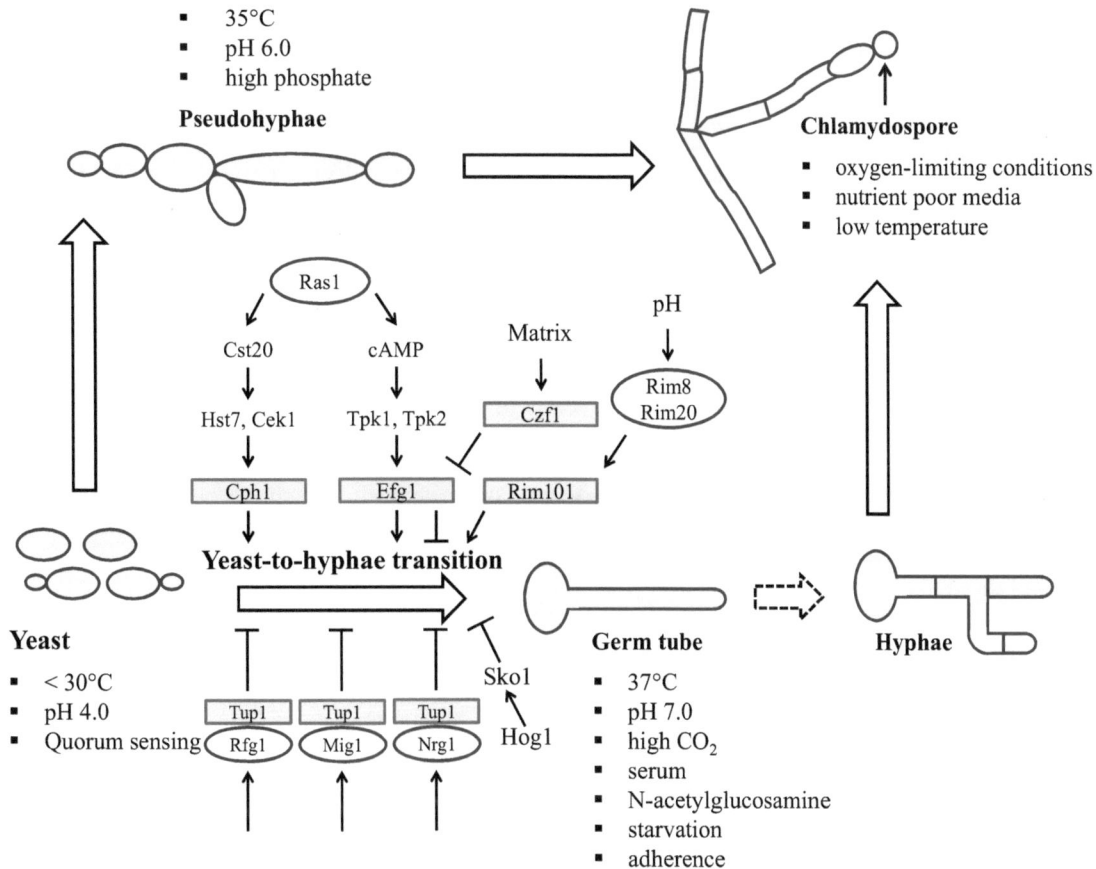

Figure 1: Morphogenesis in *C. albicans*. Induction conditions and genes regulating specific cell morphology or involved in transition process are indicated. Transcription factors are shown in gray boxes.

In addition to these morphologies, *C. albicans* can also switch between white and opaque colonies, in which the characteristic white domed colonies (containing oval cells) change into opaque flat colonies (containing oblong-shaped cells). This phenomenon is observed in the WO-1 strain and other related strains that are homozygous for the mating type-like *locus* and are able to undergo mating [33, 34]. *C. albicans* is also able to form chlamydospores, which are thick-walled cells that arise at the ends of elongated suspensor cells located on pseudohyphae or hyphae [35]. These structures are rich in lipids, RNA and carbohydrates, and are used to differentiate *Candida* species, given that only *C. albicans* and *C. dubliniensis* form chlamydospores [36]. Chlamydospores are induced under oxygen-limiting conditions, on nutrient poor media and at low temperatures [37, 38], and have rarely being observed during infection.

Regulators of Dimorphism in C. albicans

The hyphal growth is considered one of the most characteristic traits of virulent *C. albicans* strains, since it is required for tissue invasion, breaching of endothelial cells and lysis of phagocytes [39]. The yeast-to-hyphae transition is regulated by several environmental signals that include temperature, pH and compounds such as *N*-acetyl-D-glucosamine and L-proline [14, 40; Fig. **1**]. To date, several signaling pathways and regulators involved in the hyphal growth have been described [41]. Among these are the mitogen-activated protein kinase (MAPK) pathway, the cAMP-dependent protein kinase A (PKA) pathway and the pH response pathway [39]. All of them contribute to activation of transcriptional factors involved in the regulation of dimorphism in *C. albicans*.

Many of the elements of MAPK pathway in *Saccharomyces cerevisiae* also are present in *C. albicans*, these include Cst20, Hst7 and Cek1. On certain solid starvation-type (Spider or SLAD) media, mutant strains in these genes all display a defect in the hyphal morphogenesis [42, 43].

The MAPK pathway modulates the Cph1p transcriptional factor. In *C. albicans*, *Cph1p* encodes a protein that is homologous to *S. cerevisiae* Ste12p; both are activated through the pheromone response mitogen-activated protein kinase cascade, and are involved in the transcriptional activation of filamentous growth [44]. The *cph1p* mutant is partially defective in filamentous growth and shows a virulence pattern that is similar to that of the wild-type strain [45]. In contrast, a double *cph1 efg1* mutant is avirulent, and this seems to be due not only to the inability to form hyphae, but also to a loss in the expression of additional virulence genes [46].

On the other hand, the signaling pathway based on cAMP and PKA regulates the Efg1p transcription factor. The *EFG1* gene encodes a basic helix-loop-helix (bHLH) protein, which is considered an activator and repressor of morphogenesis in *C. albicans*. This dual role is controlled by environmental cues; *e.g.* in an *efg1* mutant there is no hyphal development when grown in media supplemented with serum or *N*-acetyl-D-glucosamine [47], whereas in microaerophilic/embedded conditions, mycelial growth appears to be stimulated, thus indicating that Efg1p represses filamentation under this condition [48, 49]. In addition, *efg1* mutants are unable to form chlamydospores [49]. It is known that Efg1p functions downstream of PKA and induces hypha-specific genes such as *HWP1*, *HYR1*, *ALS3*, and *ALS8*, which encode hyphal wall proteins [50, 51].

There is evidence of an Efg1p-independent pathway of filamentation in *C. albicans*, which operates under embedded/microaerophilic conditions [40]. In this sense, the transcription factor Czf1 has been suggested as an important element of this pathway. In *C. albicans* this protein is encoded by a single copy gene, of which a homologoue in *S. cerevisiae* has not been identified [52, 53]. Vinces *et al.* [52] reported that the expression of this gene is strongly induced by the carbon source, temperature, and growth phase. Moreover, Egf1 induces *CZF1*, and like *EGF1*, it is also autoregulated. Because of this, it has been suggested that *CZF1* expression and inhibition would contribute to the regulation of the activity of Efg1p during morphogenesis [52].

In addition to the cAMP, Egf1 is also regulated by the Ras1 protein whose role in filamentation has been documented. Homozygous *ras1* mutants exhibit a reduced growth rate and defects in hyphae formation under inducing conditions (*i.e.*, serum and 37 °C). $RAS1^{V13}$, a dominant-active version of Ras1p stimulates hyphal growth, while $RAS1^{Ala16}$ a dominant-negative version represses it, thus demonstrating a role of this protein in the control of filamentation [54]. Since in a double mutant *cph1/cph1 efg1/efg1*, $Ras1^{v13}$ can activate hyphal formation, it has been suggested that *C. albicans* possesses additional routes for the activation of this process [55].

Recently, the Sko1 protein has been reported as a negative regulator of the yeast-to-hyphae transition, since the deletion of *SKO1* results in an increased ability to filament compared to wild type cells, as well as an increased expression of the hypha-specific genes *ECE1* and *HWP1* without significant changes in the virulence. In addition, the yeast-to-hypha transition in a double *sko1 hog1* mutant was independent of environmental conditions (*i.e.*, pH and serum). The Sko1 orthologue in *S. cerevisiae* is controlled by the cAMP/PKA pathway, whereas in *C. albicans* Sko1 acts independently of these pathways. Besides, Sko1 acts as an important mediator in the response to oxidative stress [56].

Another important regulator of hyphal development is the *Ume6* gene, which encodes a protein with a zinc-finger DNA-binding domain that shares 41% identity with the *S. cerevisiae* Ume6. In *C. albicans*, the effect of Ume6 is dosage-dependent, since *ume6 ume6* homozygous mutant forms filaments significantly shorter in length as compared with the wild type strain and exhibits attenuated virulence; while the *ume6/UME6* heterozygous strain shows a milder filamentation defect [57]. Ume6 drives the hyphal extension *via* activation of the Hgc1 pathway [39]. The *HGC1* gene encodes a G1 cyclin-related protein, which forms a complex with CaCdc28 (cyclin-dependent kinase). This complex phosphorylates the Sep7 septin and Efg1p causing a promotion of hyphal growth [58]. During the cell cycle progression in hyphal growth, *Hgc1* transcription is dynamically regulated, reaching its highest expression in the G1 phase, due to that it has been suggested that cells are primed to form hyphae during G1 [59, 60].

In a recent study, Elson and coworkers [61] found that *C. albicans* possesses a She3-mediated system that transports specific transcripts, many of which have roles in hyphal development, into both daughter cells of budding yeast and the tip cells of hyphae,. This was demonstrated through the deletion of *She3*; which results in a reduced ability to form hyphae or production of filaments morphologically abnormal [61].

Additionally, negative regulators of hyphal development have also been identified in *C. albicans*. One of them is the transcriptional regulator Tup1 that in association with Nrg1 and Rfg1 regulate the repression of hyphal development [62, 63]. Deletion of *TUP1* causes hyperfilamentation under yeast growth conditions [62, 64]. Many of the genes repressed by Tup1 have been identified, these include genes that are induced during the yeast-to-hypha transition such as *HWP1*, *RBT1*, *RBT5*, and *WAP1*, and genes involved in pathogenesis like *RBT1* and *RBT5* [65]. The regulatory function of Tup1 is mediated by Nrg1 (a DNA-binding protein) which acts by recruiting Tup1 to target genes [63]. In addition Nrg1 represses filament-specific genes under non-inducing conditions.

Rfg1 is another protein involved in the recruitment of Tup1. Deletion of *Rfg1* results in a filamentous growth similar to *tup1* mutants. Nevertheless, *rfg1/rfg1* mutants form large amounts of true hyphae [66]. Although Rfg1 is similar in the HMG domain of the S. *cerevisiae* Rox1 protein, it does not appear to play a role in the regulation of hypoxic genes in *C. albicans* [67].

Together, these reports show that yeast-to-hyphae transition is a process that involves the participation of a complex signaling network, whose members respond to a wide variety of environmental cues.

Polarized Growth and Cell Wall Formation in C. albicans

Fungal morphogenesis is the result of polarized cell growth, in which an asymmetrical arrangement of many cellular constituents is required. In this process the cytoskeleton (actin) and polarized secretion play fundamental roles [29, 41, 68]. In particular the Rho GTPase Cdc42 and its GDP-GTP exchange factor Cdc24 are necessary for polarized growth during budding and hyphal growth in *C. albicans* and other fungi [69, 70]. This complex is known to be recruited to the germ tube and hyphal tips.

Cell wall biosynthesis has received special attention because of its importance in morphogenesis [71]. Chitin, a major structural component of fungal cell walls is made of *N*-acetyl-D-glucosamine joined by β-1,4-linkages. This linear polymer is synthesized by a transglycosylation reaction catalyzed by chitin synthases (Chs). It is known that Chss are accumulated in the cytosol of fungi, including *C. albicans*, in specialized microvesicles (chitosomes) that transport the enzyme from its site of synthesis to the site of action [72]. Various *CHS* genes have been described in *C. albicans*, i.e., *CHS1*, *CHS2, CHS3* and *CHS8* [73-76]. *CHS2* is a hyphal specific enzyme; however, it is not essential for growth, dimorphism or virulence, since *chs2* null mutants are still able to form germ tubes albeit with a reduced chitin content [74]. *CHS3* gene expression is increased during yeast-to-hyphae transtition [77]. Interestingly, *ch3* null mutants contain a reduced chitin content, aprox. 80% in hyphae and yeast compared to the wild type, but nevertheless they are not affected in virulence [75]. *CHS1* appears to be involved in septum formation, and it is essential for cell integrity and virulence [75]. The gene *CHS8* is expressed in yeast and hyphal cells. *chs8* homozygous mutants have normal growth rate, morphologies and chitin contents, although chitin synthase activity is reduced 25%. This mutant is hypersensitive to Calcofluor White [76]. Similar to *chs8* mutant, a *chs2-chs8* double mutant also has reduced chitin synthase activity (less than 3% of normal activity), and is not affected in growth or morphology, but the contents of cell wall glucan and mannan are not altered [76]. *C. albicans CHS* genes appear to have no relevance in the yeast-to-hyphae transition, however virulence is considerably affected in some *chs* mutants. This has led to propose them as adequate targets for the control of fungal infections using Chs inhibitors [71].

The genes involved in the synthesis of β-1,6-glucan, which is an abundant polysaccharide in fungal cell walls, were observed to be differentially regulated in yeast and mycelial cells. For example, the expression of *KRE6* was higher than that of *SKN1* in the yeast phase, while *SKN1* expression was strongly up-regulated upon induction of hyphae formation [78]. Homozygous *skn1* mutants were not affected in β-1,6-glucan nor β-1,3-glucan content, nor in the efficiency to form hyphae. Conversely, heterozygous *kre6* mutants had reduced by more than 80% the levels of β-1,6-glucan without affecting β-1,3-glucan levels. *kre6* homozygous mutants were not recovered, suggesting an essential role of Kre6. This was confirmed by repressing *KRE6* expression levels with a controllable gene expression system based on the *HEX1* promoter. Under this condition, *C. albicans* did not separate completely after mitosis, suggesting that optimal levels of β-1,6-glucan are required for budding completion. These cells were also susceptible to Calcofluor White [78]. On the other hand, *kre5* homozygous mutant, are unable to form hyphae on solid or liquid media, even in the presence of serum. This mutant shows reduced β-1,6-glucan levels and severe cell wall defects, and is completely avirulent [79].

Another of the genes induced during the yeast-to hyphae transition is *HWP1* whose induction is dependent on Egf1 transcription factor; it codifies a glycosyl-phosphatidylinositol-linked cell surface protein [80]. In an *HWP1* heterozygous mutant the ability to form hyphae on solid media is severely reduced and eliminated in the null mutant, but in the presence of serum, the *hwp1* null mutant is able to produce peripheral hyphae, but at reduced levels compared to the wild type strain. Because of that it has been considered that deletion of *HWP1* causes a medium-conditional defect in hyphal development [80].

In addition to aforementioned genes, several genes encoding cell wall components in *C. albicans* have been deleted generating mutants that, in most cases, had no significance on the yeast-to-hypha transition, although in some cases adherence or pathogenicity were compromised [review by 71].

pH and Morphogenesis in C. albicans

pH, in addition to temperature, is one of the most important environmental signals that determinate cell morphology in *C. albicans*. As aforementioned, growth of the yeast form is favored by temperatures below 30 °C and acidic pH, pseudohyphal growth at 35 °C and pH~6.0, and hyphal development at 37 °C and neutral pH [29, 30, 32]. Several pH-regulated genes have been identified in *C. albicans* that modulate cell morphology, cell wall architecture and adhesion properties. Among them two alkaline responsive pathways have been described, one controlled by the zinc finger transcription factor *RIM101* and the other one which is independent of this factor [20]. The Rim101/pacC pathway is conserved in several fungal species including *S. cerevisiae, U. maydis* and *Yarrowia lipolytica* [29, 81, 82]. Rim101 functionality is dependent of proteolysis, an event mediated by Rim8 and Rim20 [83]. The activation of Rim101 results in the induction of several genes such as *PRA1* and *PHR1* both required for growth and filamentation at alkaline pH [84, 85]. It has been reported that mutant strains in Rim101 pathway genes (*e.g. RIM101, RIM8, RIM13, RIM20*) exhibit little or no hyphae production at pH 7.3-8.0, as well as reduced virulence in corneal infections [29, 86].

PHR1 gene encodes a cell wall glycoprotein anchored to the membrane by glycosyl-phosphatidylinositol. This gene is highly expressed at alkaline pH [85]. Homozygous deletion mutants are affected in apical growth, being unable to form normal hyphal (germ tubes) at alkaline pH. In addition, *phr1* mutants lost their ability to cause systemic disease in mouse models [87], but are virulent in vaginal infections (pH 4.5) [88]. Another pH regulated gene is *PHR2*, a *PHR1* homologue that is not usually expressed at alkaline pH, is repressed by the Rim101 pathway and is essential for vaginal infections [83, 88]. Furthermore, Bensen *et al.* (2004) found that many pH regulated genes function in iron metabolism, and are regulated by the Rim101 pathway with a possible role in adaptation to iron starvation under alkaline conditions [89].

On the other hand, Mds3 is a Rim101p-independent regulator of the *C. albicans* alkaline response that is also necessary for filamentation at alkaline pH and virulence [20]. In addition, it is involved in chlamydospore and biofilm formation [21, 37]. Mds3 is required for the expression of *HWP1* and *ECE1* genes involved in mycelial development under alkaline pH. Recently, Zacchi *et al.* [90] found that Mds3 regulates *C. albicans* morphogenesis through the negative regulation of the TOR pathway, which senses nutrient levels. This finding links starvation and pH with morphogenesis in Candida [90].

All these studies support the concept that pH governs *C. albicans* gene expression, cellular differentiation and morphology both *in vivo* and *in vitro*.

Polyamines and Dimorphism

Polyamines (PAs) are small aliphatic amines, required for cell growth and differentiation in most organisms, and are essential for life. The most widely distributed PAs are putrescine (Put), spermidine (Spd) and spermine (Spm). PAs are involved in important cellular and physiological processes such as chromatin organization, DNA replication, RNA stabilization, transcription and translation, cell proliferation and cell death [91]. In fungi, they take part in multiple morphogenetic processes, for example, in spore germination, sporulation, yeast-to-hyphae transition and germ tube emission [92-98]. PA biosynthesis in fungi, and also in mammals, occurs only through the ornithine decarboxylase (ODC; E.C. 4.1.1.17) pathway, in which ornithine is directly transformed into Put. The higher PAs, Spd and Spm, are synthesized from Put by the action of Spd (SPDS; E.C. 2.5.1.16) and Spm synthases (E.C. 2.5.1.22) respectively, through the addition of aminopropyl groups. The aminopropyl moiety is derived from methionine, which is first converted into S-adenosylmethionine (Sam) and then decarboxylated *via* S-adenosylmethionine decarboxylase (SAMDC; E.C. 4.1.1.50). Differentiation processes in fungi are preceded by transient increases in ODC activity and in PA pools. In particular, the yeast-to-hyphae transition in dimorphic fungi has been documented for *Mucor racemosus*, *M. rouxii*, *M. bacilliformis* and *M. circinelloides* [92-94], *Yarrowia lipolytica* [95] and *C. albicans* [97], among others. It has been also described that increases in fungal ODC activity are generally not accompanied by a rise in transcript levels, suggesting that ODC activity is mostly regulated at the post-transcriptional/post-translational levels [97, 99].

PAs play a key role in the differentiation of *C. albicans*, in particular in dimorphism, where the *CaODC* gene has been characterized as essential for mycelial development [100]. The *CaODC* gene is single copy and encodes a 470 aa polypeptide, that maintains a high degree of conservation with other fungal ODC's at the central region of the protein [97]. In the presence of *N*-acetyl-D-glucosamine, an inductor of the yeast-to-hyphae transition, it has been observed that CaODC activity increases after 1 h of incubation with the inducer and persists for at least 4 h [97]. Notably, the increase in CaODC activity is not correlated with the transcript levels that remain almost constant in the yeast and hyphal forms, suggesting a post-translational mechanism of regulation of ODC activity [97]. Furthermore, inhibition of CaODC activity using concentrations of 25-50 mM of 1,4-diaminobutanone (DAB), blocks hyphal development without affecting *C. albicans* growth [101]. Homozygous *C. albicans* null *odc* mutants behave as PA auxotrophs, are unable to form mycelium in the presence of serum or other inducers, and grow exclusively in the yeast form in presence of low Put levels (0.01 mM). A notable increase in Put levels (10 mM) is necessary to restore the yeast-to-hyphae transition capacity [100]. Interestingly, these null *odc* mutant were still pathogenic in a mouse model, suggesting that somehow they must obtain polyamines from the host. Ueno *et al.* [102] have proposed that PAs act upstream the adenylate cyclase-cAMP signal pathway in *C. albicans*, showing that low PA levels due to DAB inhibition of ODC activity lead to a significant inhibition of *CYR1* mRNA levels and cAMP content. Under this condition, expression levels of *ALS1*, *ALS3*, *ALS8* genes involved in hyphae development were also reduced. These data are interesting an open the possibility that PAs regulate hyphae formation through the adenylate cyclase-cAMP pathway [102].

CONCLUSION

From the data here presented it can be concluded that *C. albicans* constitutes an excellent model for the study of dimorphism and pathogenicity. Up to know many genes and signaling pathways have been identified in the yeast-to-hypha transition processes in this fungus, nevertheless information on new molecules are still needed to fully understand the mechanisms governing morphogenesis.

REFERENCES

[1] Kim J, Sudbery P. *Candida albicans*, a major human fungal pathogen. J Microbiol 2011; 49(2): 171-177.

[2] Edmond MB, Wallace SE, McClish DK, Pfaller MA, Jones RN, Wenzel RP. Nosocomial bloodstream infections in United States hospitals: a three year analysis. Clin Infect Dis 1999; 29: 239-244.

[3] d'Enfert C. Biofilms and their role in the resistance of pathogenic *Candida* to antifungal agents. Curr Drug Targets 2006; 7: 465-470.

[4] ten Cate JM, Klis FM, Pereira-Cenci T, Crielaard W, de Groot PWJ. Molecular and cellular mechanisms that lead to *Candida* biofilm formation. J Dent Res 2009; 88(2): 105-115.

[5] Klepser ME. *Candida* resistance and its clinical relevance. Pharmacotherapy 2006; 26: 68S-75S.

[6] Berman J. Morphogenesis and cell cycle progression in *Candida albicans*. Curr Opin Microbiol 2006; 9: 595-601.

[7] Kibbler CC, Seaton S, Barnes RA, Gransden WR, Holliman RE, Johnson EM, Perry JD, Sullivan DJ, Wilson JA. Management and outcome of bloodstream infections due to *Candida* species in England and Wales. J Hosp Infect 2003; 54: 18-24.

[8] Fitzpatrick DA, Logue ME, Stajich JE, Butler G. A fungal phylogeny based on 42 complete genomes derived from supertree and combined gene analysis. BMC Evol Biol 2006; 6: 99.

[9] Jones T, Federspiel NA, Chibana H, Dungan J, Kalman S, *et al.* The diploid genome sequence of *Candida albicans*. PNAS 2004; 101: 7329-7334.

[10] Butler G, Rasmussen MD, Lin MF, Santos MAS, *et al.* Evolution of pathogenicity and sexual reproduction in eight *Candida* genomes. Nature 2009; 459(4): 657-662.

[11] Braun BR, van het Hoog M, d'Enfert C, Martchenko M, *et al.* A human-curated annotation of the *Candida albicans* genome. Plos Genet 2005;1(1): e1.

[12] Santos MAS, Tuite MF. The CUG codon is decoded *in vivo* as serine and not leucine in *Candida albicans*. Nucl Acids Res 1995; 23(9): 1481-1486.

[13] Enloe B, Diamond A, Mitchell AP. A single-transformation gene function test in diploid *Candida albicans*. J Bacteriol 2000; 182: 5730-5736.

[14] Molero G, Díez-Orejas R, Navarro-García F, Monteoliva L, Pla J, Gil C, Sánchez-Pérez M, and Nombela C. *Candida albicans*: genetics, dimorphism and pathogenicity. Internat Microbiol 1998; 1: 95-106.

[15] Noble SM, Johnson AD. Genetics of *Candida albicans* a diploid human fungal pathogen. Annu Rev Genet 2007; 41: 193-211.

[16] Fonzi WA, Irwin MY. Isogenic strain construction and gene mapping in *Candida albicans*. Genetics 1993; 134: 717-728.

[17] Roemer T, Jiang B, Davison J, Ketela T, Veillette K. Large-scale essential gene identification in *Candida albicans* and applications to antifungal drug discovery. Mol Microbiol 2003; 50(1): 167-181.

[18] Nakayama H, Mio T, Nagahashi S, Kokado M, Arisawa M, Aoki Y. Tetracycline-regulatable system to tightly control gene expression in the pathogenic fungus *Candida albicans*. Infect Immun 2000; 68: 6712-6719.

[19] Noble SM, Johnson AD. Strains and strategies for large-scale gene deletion studies of the diploid human fungal pathogen *Candida albicans*. Eukaryot Cell 2005; 4: 298-309.

[20] Davis DA, Bruno VM, Loza L, Filler SG, Mitchell AP. *Candida albicans* Mds3p, a conserved regulator of pH responses and virulence identified through insertional mutagenesis. Genetics 2002; 162: 1573-1581.

[21] Richard ML, Nobile CJ, Bruno VM, Mitchell AP. *Candida albicans* biofilm-defective mutants. Eukaryot Cell 2005; 4: 1493-1502.

[22] Norice CT, Smith FJ, Solis N, Filler SG, Mitchell AP. Requirement for Candida albicans Sun41 in biofilm formation and virulence. Eukaryot Cell 2007; 6(11): 2046-2055.

[23] Lavoie H, Sellam A, Askew C, Nantel A, Whiteway M. A toolbox for epitope-tagging and genome-wide location analysis in *Candida albicans*. BMC Genomics 2008; 9: 578.

[24] Hogues H, Lavoie H, Sellam A, Mangos M, Roemer T, Purisima E, Nantel A, Whiteway M. Transcription factor substitution during evolution of fungal ribosome regulation. Mol Cell 2008; 29: 552-562.

[25] Ruiz-Herrera J, Sentandreu R. Different effectors of dimorphism in *Yarrowia lipolytica*. Arch Microbiol 2002; 178: 477-483.

[26] Rappleye CA, Goldman WE. Defining virulence genes in the dimorphic fungi. Annu Rev Microbiol 2006; 60: 281-303.

[27]   Nadal M, García-Pedrajas MD, Gold SE. Dimorphism in fungal plant pathogens. FEMS Microbiol Lett 2008; 284: 127-134.

[28]   Klose J, De Sá MM, Kronstad JW. Lipid-induced filamentous growth in *Ustilago maydis*. Mol Microbiol 2004; 52(3): 823-835.

[29]   Biswas S, Dijck PV, and Datta A. Environmental sensing and signal transduction pathways regulating morphopathogenic determinants of *Candida albicans*. Microbiol Mol Biol Rev 2007; 71(2): 348-376.

[30]   Nickerson KW, Atkin AL, Hornby JM. Quorum sensing in dimorphic fungi: farnesol and beyond. Appl Environ Microbiol 2006; 72(6): 3805-3813.

[31]   Rooney PJ, Klein BS. Linking fungal morphogenesis with virulence. Cell Microbiol 2002; 4: 127-137.

[32]   Sudbery P, Gow N, Berman J. The distinct morphogenic states of *Candida albicans*. Trends Microbiol 2004; 12(7): 317-324.

[33]   Miller MG, Johnson AD. White-opaque switching in *Candida albicans* is controlled by mating-type locus homeodomain proteins and allows efficient mating. Cell 2002; 110: 293-302.

[34]   Slutsky B, Staebell M, Anderson J, Risen L, Pfaller M, Soll DR. White-opaque transition: a second high frequency transition in *Candida albicans*. J Bacteriol 1987; 169: 189-197.

[35]   Odds FC. Candida and candidosis, 2nd ed. London, UK: Bailliere Tindall. 1988

[36]   Staib P, Morschhauser J. Chlamydospore formation on Staib agar as a species specific characteristic of *Candida dubliniensis*. Mycoses 1999; 42: 521-24.

[37]   Nobile CJ, Bruno VM, Richard ML, Davis DA, Mitchell AP. Genetic control of chlamydospore formation in *Candida albicans*. Microbiology 2003; 149: 3629-3637.

[38]   Staib P, Morschäuer J. Chlamydospore formation in *Candida albicans* and *Candida dubliniensis* - an enigmatic developmental programme. Mycoses 2006; 50: 1-12.

[39]   Carlisle PL and Kadosh D. *Candida albicans* Ume6, a filament-specific transcriptional regulator, directs hyphal growth *via* a pathway involving Hgc1 cyclin-related protein. Eukaryotic Cell 2010; 9(9): 1320-1328.

[40]   Ernst JF. Transcription factors in *Candida albicans*-environmental control of morphogenesis. Microbiology 2000; 146: 1763-1774.

[41]   Liu H. Transcriptional control of dimorphism in *Candida albicans*. Curr Opin Microbiol 2001; 4: 728-735.

[42]   Csank C, Schroppel K, Leberer E, Harcus D, Mohamed O, Meloche S, Thomas DY, Whiteway M. Roles of the *Candida albicans* mitogen-activated protein kinase homolog, Cek1p, in hyphal development and systemic candidiasis. Infect Immun 1998; 66: 2713-2721.

[43]   Leberer E, Harcus D, Broadbent ID, Clark KL, Dignar D, Ziegelbeuer K, Schmidt A, Gow HAR, Brown AJP, Thomas DY. Signal transduction through homologs of the Ste20p and Ste7p protein kinases can trigger hyphal formation in the pathogen fungus *Candida albicans*. Proc Natl Acad Sci USA 1996; 93: 13217-13222.

[44]   Liu H, Köhler J, and Fink GR. Suppression of hyphal formation in *Candida albicans* by mutation of a *STE12* homolog. Science 1994; 266: 1723-1725.

[45]   Lo HJ, Köhler JR, DiDomenico B, Loebenberg D, Cacciapuoti A, and Fink GR. Nonfilamentous *C. albicans* mutants are avirulent. Cell 1997; 90: 939-49.

[46]   Staib P, Kretschmar M, Nichterlein T, Hof H, and Morschhäuser J. Transcriptional regulators Cph1p and Efg1p mediate activation of the *Candida albicans* virulence gene *SAP5* during infection. Infect Immun 2002; 70(2): 921-927.

[47]   Stoldt VR, Sonneborn A, Leuker CE, and Ernst JF. Efg1p, an essential regulator of morphogenesis of the human pathogen *Candida albicans*, is a member of a conserved class of bHLH proteins regulating morphogenetic processes in fungi. EMBO J 1997; 16(8): 1982-1991.

[48]   Riggle PJ, Andrutis KA; Chen X, Tzipori SR, and Kumamoto CA. Invasive lesions containing filamentous forms produced by a *Candida albicans* mutant that is defective in filamentous growth in culture. Infect Immun 1999; 67(7): 3649-3652.

[49]   Sonneborn A, Bockmühl DP, and Ernst JF. Chlamydospore Formation in *Candida albicans* requires the Efg1p morphogenetic regulator. Infect Immun 1999; 67(10): 5514-5517.

[50]   Kumamoto CA and Vinces MD. Contributions of hyphae and hypha-co-regulated genes to *Candida albicans* virulence. Cell Microbiol 2005; 7(11): 1546-1554.

[51]   Bailey DA, Feldmann PJF, Bovey M, Gow NAR, and Brown AJP. The *Candida albicans HYR1* gene, which is activated in response to hyphal development, belongs to a gene family encoding yeast cell wall proteins. J Bacteriol 1996; 178(18): 5353-5360.

[52]   Vinces MD, Haas C, Kumamoto CA. Expression of the *Candida albicans* morphogenesis regulator gene CZF1 and its regulation by Efg1p and Czf1p. Eukaryot Cell 2006; 5(5): 825-835.

[53]   Whiteway M. Transcriptional control of cell type and morphogenesis in *Candida albicans*. Curr Opin Microbiol 2000; 3: 582-588.

[54]   Feng Q, Summers E, Guo B, and Fink G. Ras signaling is required for serum-induced hyphal differentiation in *Candida albicans*. J Bacteriol 1999; 181(20): 6339-6346.

[55]   Chen J, Zhou S, Wang Q, Chen X, Pan T, and Liu H. Crk1, a novel Cdc2-related protein kinase, is required for hyphal development and virulence in *Candida albicans*. Mol Cell Biol 2000; 20(23): 8696-8708.

[56]   Alonso-Monge R, Román E, Arana DM, Prieto D, Urrialde V, Nombela C, Pla J. The Sko1 protein represses the yeast-to-hypha transition and regulates the oxidative stress response in *Candida albicans*. Fungal Genet Biol 2010; 47: 587-601.

[57]   Banerjee M, Thompson DS, Lazzell A, Carlisle PL, Pierce C, Monteagudo C, López-Ribot JL, and Kadosh D. UME6, a novel filament-specific regulator of *Candida albicans* hyphal extension and virulence. Mol Biol Cell 2008; 19: 1354-1365.

[58]   Zheng X, Wang Y, and Wang Y. Hgc1, a novel hypha-specific G1 cyclin-related protein regulates *Candida albicans* hyphal morphogenesis. EMBO J 2004; 23(8): 1845-1856.

[59]   Wangn A, Lane S, Tian Z, Sharon A, Hazan I, and Liu H. Temporal and spatial control of *HGC1* expression results in Hgc1 localization to the apical cells of hyphae in *Candida albicans*. Eukaryot Cell 2007; 6(2): 253-261.

[60]   Whiteway M, Bachewich C. Morphogenesis in *Candida albicans*. Annu Rev Microbiol 2007; 61: 529-553.

[61]   Elson SL, Noble SM, Solis NM, Filler SG, Johnson AD. An RNA transport system in *Candida albicans* regulates hyphal morphology and invasive growth. PLoS Genet 2009; 5(9).

[62]   Braun BR, Johnson AD. Control of filament formation in *Candida albicans* by the transcriptional repressor. Science 1997; 277(5322): 105-109.

[63]   Braun BR, Kadosh D, and Johnson AD. NRG1, a repressor of filamentous growth in *C.albicans*, is down-regulated during filament induction. EMBO J 2001; 20(17): 4753-4761.

[64]   Braun BR, Johnson AD. *TUP1*, *CPH1* and *EFG1* make independent contributions to filamentation in *Candida albicans*. Genetics 2000; 155: 57-67.

[65]   Braun BR, Head WS, Wang MX, and Johnson AD. Identification and characterization of TUP1-regulated genes in *Candida albicans*. Genetics 2001; 156: 31-44.

[66]   Khalaf RA, Zitomer RS. The DNA binding protein Rfg1 is a repressor of filamentation in *Candida albicans*. Genetics 2001; 157: 1503-1512.

[67]   Kadosh D, Johnson AD. Rfg1, a protein related to the *Saccharomyces cerevisiae* hypoxic regulator Rox1, controls filamentous growth and virulence in *Candida albicans*. Mol Cell Biol 2001; 21(7): 2496-2505.

[68]   Akashi T, Kanbe T, Tanaka K. The role of the cytoskeleton in the polarized growth of the germ tube in *Candida albicans*. Microbiology 1994; 140: 271-280.

[69]   Ushinsky SC, Harcus D, Ash J, Dignard D, Marcil A, Morchhauser J, Thomas DY, Whiteway M, Leberer E. CDC42 is required for polarized growth in human pathogen *Candida albicans*. Eukaryot Cell 2002; 1: 95-10.

[70]   Bassilana M, Hopkins J, Arkowitz RA. Regulation of the Cdc42/Cdc24 GTPase module during *Candida albicans* hyphal growth. Eukaryot Cell 2005; 4(3): 588-603.

[71]   Ruiz-Herrera J, Elorza MV, Valentín E, Sentandreu R. Molecular organization of the cell wall of *Candida albicans* and its relation to pathogenicity. FEMS Yeast Res 2006; 6: 14-29.

[72]   Ruiz-Herrera J, Elorza MV, Alvarez PE, Sentandreu R. Synthesis of the fungal cell wall. Pathogenic Fungi: Structural Biology and Taxonomy (San-Blas G & Calderone R, eds). Caister Academic Press, Wymondham, UK. 2004; 41-99.

[73]   Chen-Wu JL, Zwicker J, Bowen AR, Robbins PW. Expression of chitin synthase genes during yeast and hyphal growth phases of *Candida albicans*. Mol Microbiol 1992; 6(4): 497-502.

[74]   Gow NAR, Robbins PW, Lester JW, Brown AJP, Fonzi WA, Chapman T, Kinsman S. A hyphal-specific chitin synthase gene (CHS2) is not essential for growth, dimorphism, or virulence of *Candida albicans*. PNAS 1994; 91: 6216-6220.

[75]   Mio T, Yabe T, Sudoh M, Satoh Y, Nakajima T, Arisawa M, Yamada-Okabe H. Role of three chitin synthase genes in the growth of *Candida albicans*. J Bacteriol 1996; 178(8): 2416-2419.

[76]   Munro CA, Whitton RK, Hughes HB, Rella M, Selvaggini S, Gow NAR. *CHS8*—a fourth chitin synthase gene of *Candida albicans* contributes to *in vitro* chitin synthase activity, but is dispensable for growth. Fungal Genet Biol 2003; 40(2): 146-158.

[77]   Sudo M, Nagahashi S, Doi M, Ohta A, Takagu M, Arisawa M. Cloning of the chitin synthase 3 gene from *Candida albicans* and its expression during yeast-hyphal transition. Mol Gen Genetics 1993; 241; 351-358.

[78]   Mio T, Yamada-Okabe T, Yabe T, Nakajima T, Arisawa M, Yamada-Okabe H. Isolation of the *Candida albicans* homologs of *Saccharomyces cerevisiae* KRE6 and SKN1: expression and physiological function. J Bacteriol 1997; 179(7): 2363-2372.

[79]   Herrero AB, Magnelli P, Mansour MK, Levitz SM, Bussey H, Abeijon C. *KRE5* gene null mutant strains of *Candida albicans* are avirulent and have altered cell wall composition and hypha formation properties. Eukaryot Cell 2004; 3(6):1423-1432.

[80]   Sharkey LL, McNemar MD, Saporito-Irwin SM, Sypherd PS, Fonzi WA. *HWP1* functions in the morphological development of *Candida albicans* downstream of EFG1, TUP1, and RBF1. J Bacteriol 1999; 181: 5273-5279.

[81]   Lambert M, Blanchin-Roland S, Le Louedec F, Lépingle A, Gaillardin C. Genetic analysis of regulatory mutants affecting synthesis of extracellular proteinases in the yeast *Yarrowia lipolytica*: identification of a *RIM101/pacC* homolog. Mol Cell Biol 1997; 17(7): 3966-3976.

[82]   Aréchiga-Carvajal ET, Ruiz-Herrera J. The *RIM101/pacC* homologue from the Basidiomycete *Ustilago maydis* is functional in multiple pH-sensitive phenomena. Eukaryot Cell 2005; 4(6): 999-1008.

[83]   Davis D, Wilson RB, Mitchell AP. *RIM101*-dependent and -independent pathways govern pH responses in *Candida albicans*. Mol Cell Biol 2000; 20(3): 971-978.

[84]   Sentandreu M, Elorza MV, Sentandreu R, and Fonzi WA. Cloning and characterization of *PRA1*, a gene encoding a novel pH-regulated antigen of *Candida albicans*. J Bacteriol 1998; 180(2): 282-289.

[85]   Saporito-Irwin SM, Birse CE, Sypherd PS, Fonzi WA. *PHR1*, a pH-regulated gene of *Candida albicans*, is required for morphogenesis. Mol Cell Biol 1995; 15(2): 601-613.

[86]   Yuan X, Mitchell BM, Hua X, Davis DA, and Wilhelmus KR. The *RIM101* signal transduction pathway regulates *Candida albicans* virulence during experimental keratomycosis. Invest Ophthalmol Vis Sci 2010; 51(9): 4668-4676.

[87]   Ghannoum MA, Spellberg B, Saporito-Irwin SM, Fonzi WA. Reduced virulence of *Candida albicans PHR1* mutants. Infect Immun 1995; 63: 4528-4530.

[88]   De Bernardis F, Muhlschlegel FA, Cassone A, Fonzi WA. The pH of the host niche controls gene expression in and virulence of *Candida albicans*. Infect Immun 1998; 66(7): 3317-3325.

[89]   Bensen ES, Martin SJ, Li M, Berman J, Davis DA. Transcriptional profiling in *Candida albicans* reveals new adaptive responses to extracellular pH and functions for Rim101. Mol Microbiol 2004; 54(5): 1335-1351.

[90]   Zacchi LF, Gomez-Raja J, Davis DA. Mds3 Regulates Morphogenesis in *Candida albicans* through the TOR Pathway. Mol Cell Biol 2010; 30(14): 3695-3710.

[91]   Childs AC, Mehta DJ, Gerner EW. Polyamine-dependent gene expression. Cell Mol Life Sci 2003; 60: 1394-1406.

[92]   Calvo-Méndez C, Martínez-Pacheco M, Ruiz-Herrera J. Regulation of ornithine decarboxylase activity in *Mucor bacilliformis* and *Mucor rouxii*. Exp Mycol 1987; 11: 270-277.

[93]   Inderlied CB, Cihlar RL, Sypherd PS. Regulation of ornithine decarboxylase during morphogenesis of *Mucor racemosus*. J Bacteriol 1980; 141: 699-706.

[94]   Blasco JL, García-Sánchez MA, Ruiz-Herrera J, Eslava AP, Iturriaga EA. A gene encoding for ornithine decarboxylase (*odcA*) is differentially expressed during the *Mucor circinelloides* yeast-to-hypha transition. Res Microbiol 2002; 153(3): 155-164.

[95]   Guevara-Olvera L, Calvo-Méndez C, Ruiz-Herrera J. The role of polyamine metabolism in dimorphism of *Yarrowia lipolytica*. J Gen Microbiol 1993; 193: 485-493.

[96]   Martinez-Pacheco M, Rodriguez G, Reyna G, Calvo-Mendez C, Ruiz-Herrera J. Inhibition of the yeast-to-hypha transition and the phorogenesis of Mucorales by diamino butanone. Arch Microbiol 1989; 151: 10-14.

[97]   López MC, García S, Ruiz-Herrera J, Domínguez A. The ornithine decarboxylase gene from *Candida albicans*. Sequence analysis and expression during dimorphism. Curr Genet 1997; 32: 108-114.

[98]   Ruiz-Herrera J. Polyamines, DNA methylation, and fungal differentiation. Crit Rev Microbiol 1994; 20: 143-150.

[99]   Jimenez-Bremont JF, Ruiz-Herrera J. Analysis of the transcriptional regulation of *YlODC* gene from the dimorphic fungus *Yarrowia lipolytica*. Microbiol Res 163(6): 717-723.

[100]   Herrero AB, Lopez MC, García S, Schmidt A, Spaltmann F, Ruiz-Herrera J, Dominguez A. Control of filament formation in *Candida albicans* by polyamine levels. Infect Immun 1999; 67(9): 4870-4878.

[101]   Martinez JP, Lopez-Ribot JL, Gil ML, Sentandreu R, Ruiz-Herrera J. Inhibition of the dimorphic transition of *Candida albicans* by the ornithine decarboxylase inhibitor 1,4-diaminobutanone: alteration in the glycoprotein composition of the cell wall. J Gen Microbiol 1990; 136: 1937-1943.

[102]   Ueno Y, Fukumatsu M, Ogasawara A, Watanabe T, Mikami T, Matsumoto T. Hyphae formation in *Candida albicans* is regulated by polyamines. Biol Pharm Bull 2004; 27(6): 890-892.

CHAPTER 4

Histoplasma capsulatum and its Virulence Determinants

Amalia Porta, Elena Calabrese, Ilaria Granata and Bruno Maresca[*]

Dept. of Pharmaceutical and Biomedical Science, University of Salerno, Italy

Abstract: The pathologist Samuel Taylor Darling discovered in 1905 a new disease caused by a previously not described microrganism that he named *Histoplasma capsulatum* based on an archaic term for macrophages (Histo), its resemblance to protozoan parasites (plasma), and the apparent presence of a surrounding capsule (capsulatum). However, it turned out that it was a fungus and not encapsulated. In the last few decades, fungal infections have become more widespread due to the AIDS epidemic end to the increase in immune compromised state of patients under chemotherapic treatment, extensive use of antibiotics and organ transplants. In parallel with the developments of molecular biology tools, our knowledge of the mechanism of virulence and host-parasite interactions have also increased significantly. This chapter will highlight the major findings in these areas of investigation focusing on the biology of the virulence determinants of *H.capsulatum*.

Keywords: Histoplasma, samuel taylor daring, virulence, phagocytic cells, conidia, yeast, hyphae, cell wall, chitin, glucan, lectin, integrin

INTRODUCTION

H. capsulatum is a dimorphic pathogenic fungus which is the causative agent of histoplasmosis, a worldwide disease. *Histoplasma* is prevalent in the soil, and is endemic in USA with 40 million people infected [1] particularly in the Ohio and Mississippi River valleys in areas contaminated with bird or bat excrement and infection occurs by inhaling aerosol containing microconidia of *H. capsulatum* into the lung alveoli. *H. capsulatum* is an intracellular fungal pathogen that causes respiratory and systemic disease by proliferating within phagocytic cells. This fungal pathogen switches reversibly from a saprophytic mould form (septate hyphae) in the soil to a parasitic, budding yeast form in a competent mammalian host. Infection occurs when the mycelial form releases microconidia (1-4 x 2-6 μm) or small mycelia fragments (5-8 μm) that, if inhaled, reach the terminal bronchioles and alveoli of the lung. In the alveolar space conidia are phagocytized by resident macrophages where conidia convert to the unicellular yeast form. Phagocytosis of microconidia and small mycelial fragments take place within 4 to 6 h, with a complete conversion to the yeast phase within 72 h. Initially, *Histoplasma* yeasts multiply inside alveolar macrophages and spread throughout the reticuloendothelial system [1]. Dendritic cells (DC) ingest and kill the yeasts and present *Histoplasma* antigen to naive T-lymphocytes [2]. Within 2 to 3 weeks, a T cell-mediated immune response is generated and is responsible for halting dissemination by assisting intracellular killing of the yeast by effector macrophages. In immune-depressed patients, *Histoplasma* proliferates and disseminates throughout the body, causing tissue destruction and progressive dissemination that can be fatal if untreated. There are no reported cases of transmission of the diseases between infected individuals. Exposure to body temperatures is necessary for conidia or hyphae to differentiate into yeast cells that are found in the infected tissues and are responsible for the pathogenesis of histoplasmosis [2]. The morphologic transition is a lengthy process when is induced *in vitro* by shifting the temperature of incubation from 25° to 37°C, whereas it is relatively rapid within the host. The phenotypic switch results in a change not only of cell shape, but also in the composition of the cell wall, the presence of antigenic molecules, expression of specific genes required for early adaptation to elevated temperatures, maintenance of the yeast state and important virulence traits. The factors that affect morphogenic process are of considerable interest, since the yeast phase is required

[*]**Address correspondence to Bruno Maresca:** Dept. of Pharmaceutical and Biomedical Science, University of Salerno, Italy, Email: bmaresca@unisa.it

for pathogenicity. In this regard, Maresca *et al.* showed that the sulfhydryl group inhibitor p-chloromercuriphenylsulfonic acid (PCMS) prevents irreversibly the mycelium to yeast transition [3].

Later, Medoff *et al.* [4] showed that the PCMS-treated mycelia when injected into a mouse model of infection did not cause the disease implying that the transition to the yeast form is an absolute requirement for progression of histoplasmosis.

H. capsulatum Cell Wall and Host Immune System

The shape, the integrity and almost every aspect of the biology (saprophytic and parasitic) of *H. capsulatum* is tightly correlated to changes of the cell wall that occur constantly during cell division, growth and morphogenesis. The major components of the *H. capsulatum* cell wall are three polysaccharides: a polymer of *N*-acetylglucosamine called chitin, β-glucans [β-(1,3)- and β-(1, 6)-linked], and α-(1,3)-glucan [5] and their concentration varies depending on culture medium composition, environmental conditions, strain type and the extraction method used. In fungi, the polysaccharide-rich cell wall is a major antigen recognized by the mammalian immune system that responds to phagocytosis, production of antimicrobial compounds, and induction of pro-inflammatory cytokines [6]. The α and β-glucans have different biological roles. [6, 7]. Although the mycelial form of *Histoplasma* lacks entirely α-(1,3)-glucan, the germination of conidia of *Histoplasma* at 37°C into yeast cells is associated with the production of this polysaccharide. Depending on the ratio of chitin to glucan in the yeast cell wall *H. capsulatum* is classified as chemotype I and II. Chemotype I contains more chitin than chemotype II in which glucan is predominantly linked in the α-configuration, while chemotype I has low level of glucan in the α-configuration and detectable levels of β-linked glucan [6]. Both chemotypes produce this polysaccharide *in vivo* but, *in vitro*, chemotype I synthesizes α-(1,3)-glucan at a non-detectable low level. However, no correlation was found between virulence and amount of α-(1,3)-glucan of these chemotypes. The first direct evidence that α-(1,3)-glucan plays a role in *Histoplasma* virulence came from silencing *AGS1* (a gene coding a α-(1,3)-glucan synthase) by RNA interference (RNAi) or deleting of *AGS1* yielding attenuation of the ability to kill macrophages and colonize murine lungs [6]. Analysis of the *AGS1 locus* of chemotype I revealed a 2.7-kb insertion in the promoter region that causes a decrease in *AGS1* expression. Nonetheless, *AGS1* mRNA is detected during respiratory infection with chemotype I yeast, confirming that α-(1,3)-glucan could be produced during *in vivo* growth despite its absence *in vitro*. Moreover, Edwards *et al.* [6] recently demonstrated that *AGS1* expression is not critical for chemotype I yeast virulence, whereas is clearly required for chemotype II strains. Chemotype I *ags1⁻* mutant strains do not affect intracellular growth of yeast during macrophage infection and in a murine model of histoplasmosis, showing no impairment of lung infection or extra-pulmonary dissemination. Despite the absence of cell wall α-(1,3)-glucan, chemotype I yeast can avoid detection by the host immune system. In addition to *AGS1*, biosynthesis of α-(1,3)-glucan in the cell wall requires the presence of the *AMY1* gene product, a homolog protein of the α-amylase family of glycosyl hydrolases, and Ugp1, a UTP-glucose-1-phosphate uridylyltransferase which synthesizes UDP-glucose monomers that are incorporated into glucan polysaccharides. Loss of *AMY1* function, reduced the ability of *Histoplasma* to kill macrophages *in vitro* and to colonize murine lungs [7]. Moreover, subculturing *H. capsulatum* extensively induce the emergence of spontaneous avirulent mutants that irreversibly did not synthesize α-(1,3)-glucan [8]. However, it has not been established whether the loss of this polysaccharide causes the decreased virulence or is indirectly due to some other genetic modification. The lack of α-(1,3)-glucan in the avirulent mutant strains may result in secondary effects of the cell wall architecture. Topographically, in the yeast cell wall, the α-(1,3)-glucan overlaps and masks the β-glucan recognition and the subsequent activation of immune responses [10]. The first step of the mammalian innate immune system during a microbial infection is the recognition of molecules [pathogen-associated molecular patterns (PAMPs)] that can be either secreted or present on the pathogen's surface. Fungi have a variety of PAMPs including β-glucan, chitin and mannoproteins that are recognized by a number of host trans-membrane pattern-recognition receptors (PRRs) such as Toll-like receptors TLR2 and TLR4, collectins SP-A and SP-D, pentraxin-3, CR3 integrin, and C-type lectins that are important to detect fungal-associated carbohydrates [9]. Dectin-1 is a type II membrane receptor that belongs to the C-type lectin family. In particular, an extracellular C-type lectin domain whose substrates are oligomers of β-(1,3)-glucans, is linked, *via* a transmembrane domain, to an intracellular signaling tail. In some isoforms a stalk region linking the C-type lectin domain to the transmembrane domain is present. Eight isoforms of dectin-1 have been cloned from human tissues, resulting from alternate splicing of the corresponding mRNA. However, both humans and mice preferentially express the B or stalkless isoform of dectin-1 [10]. β-glucan recognition initiates signaling cascades, contributing to macrophage, DC, and neutrophil responses, including phagocytosis, oxidative burst, neutrophil degranulation, fungal killing, and production of cytokines and chemokines that coordinate the recruitment and the action of immune effector cells [14]. The presence of the α-(1,3)-glucan in the outermost layer of the *H. capsulatum* yeast cell wall may contribute to pathogenesis by blocking the detection of immunostimulatory β-(1,3)-glucans by the dectin-1 receptor on host phagocytes [11]. Despite the absence of cell wall α-(1,3)-glucan, chemotype I yeast can avoid detection by dectin-1 in a growth stage-dependent manner. This suggests that the production of a *Histoplasma* chemotype I factor that may circumvents the α-(1,3)-glucan requirement for yeast virulence [11]. Detection by dectin-1 may be by-passed by other β-glucan receptors. During the early phase of infection, alveolar macrophages recognize unopsonized *Histoplasma* yeasts and microconidia through CD11/CD18 β-integrin receptors. These glycoproteins constitute the complement receptors (CR1, CR3, and CR4) on the surface of phagocytes. CR3 (CD11b/CD18 heterodimers) can interact with various set of microbial ligands, including fungal β-glucans [12, 13]. Long *et al.* [14] using purified human CR3 (CD11b/CD18) identified, from the cell wall and cell membrane of yeast cells a single 60-kDa protein with the same isoelectric point and molecular mass of *Histoplasma* heat shock protein 60 (HSP60). Although HSP60 is present predominantly in the cytosolic fraction of cells [15] is, also, expressed in clusters on the cell wall of *Histoplasma* yeasts as an immunodominant antigen, acting as a ligand that mediates the attachment to macrophage CD11/CD18. Immunization with several protein-containing antigens and extracts from *H.capsulatum*, in particular, with HSP60, induce T-cell activation and the release of Th1-associated proinflammatory cytokines, such as interleukin-12 (IL-12), gamma interferon (IFN-γ), and TNF-α [16]. There are two major mechanisms by which phagocytes kill Histoplasma: production of toxic oxygen radicals and/or NO and through phagosome-lysosome fusion (PL-fusion) with concomitant acidification, and lysosomal hydrolases activity. The phagolysosome is perhaps the most effective antimicrobial site within macrophages due both to its acidity and to its variety of hydrolytic enzymes. Phagocytosis of *H. capsulatum* yeast cells by mouse macrophages leads to normal phagolysosomal PL-fusion [17]. In mouse macrophages, *Histoplasma* yeasts survive modulating the intraphagosomal pH to about 6.5 [18]. At this pH, lysosomal acid hydrolases that enter the phagosome presumably are inactive [22]. Survival of *Histoplasma* yeasts within human macrophages requires limited PL-fusion, whereas DC exhibit marked PL-fusion, suggesting that DC lysosomal hydrolases are sufficient to kill and degrade *H. capsulatum* yeasts [19]. Interesting, once phagocytated by human macrophages, *Histoplasma* yeasts control intraphagosomal pH that remains at pH <6.4-6.5. Thus, human macrophages lysosomal hydrolases do not require an acid pH to kill *Histoplasma* yeasts [20].

Mycelium to Yeast Transition: Membrane Physical State and Heat Shock Gene Transcription

The rapid expression of members of a family of related and highly conserved proteins called heat shock proteins (HSPs) is a universal cellular response (Heat Shock Response) to a variety of environmental stresses, such as heat, drought, salinity, osmotic shock, membrane shearing, ischemia, *etc.* [21]. Some HSPs are present in unstressed cells (HSC, heat shock cognate genes) and play important roles for basic cellular processes such as protein folding, translocation, mRNA splicing, nucleic acid and protein syntheses, mitochondrial electron transport, chloroplast photosynthesis, *etc.* [22, 23]. HSPs, also named "molecular chaperones", have been classified in different families according to their molecular weight varying from low (between 8 and 32-kDa) and high molecular weight (40, 60, 70, 90 and 100-kDa) as described in Table **1**.

Table 1: Heat shock proteins

Gene Family	Protein Isoforms	Localization	Function
Ubiquitin	Ubiquitin	Cytosol	Protein degradation
Small HSPs	HSP10	Mitochondria	Tolerance to ischemia
	HSP20-HSP30	Cytosol/nucleus	Oxidative stress tolerance
Hsp30	HSP32	Cytosol	Heme oxygenase
	HSP35	Cytosol	G3PDH
Hsp40	HSP40	Cytosol, Nucleus	HSP72 co chaperone

	HSP47	Endoplasmic reticulum	Normal protein folding
Hsp60	HSP58	Mitochondria	Normal protein folding
	HSP60	Cytosol	
Hsp70	HSP72	Cytosol/nucleus	Thermotolerance
	HSP73	Cytosol/nucleus	Thermotolerance
	Grp78	Endoplasmic reticulum	Normal protein folding
	mtHSP70	Mitochondria	Protein transport to mitochondria
HSP90	HSP90	Cytosol/nucleus	Steroid hormone receptors
	Grp94	Endoplasmic reticulum	interaction
Hsp100	HSP104	Cytosol	Thermotolerance
	HSP110	Nucleus/nucleolus	
	Grp 100	Endoplasmic reticulum/Golgi	Tolerance to ischemia

It has been shown that fungal pathogens, such as *H. capsulatum*, and bacteria and protozoan parasites, at the onset of infection of macrophages and/or of other cells stimulate the transcriptional activation of two sets of genes: virulence genes (species specific) and genes involved in the heat shock response (evolutionary conserved) [24]. Virulence genes, induced during host invasion operate in a coordinate fashion and allow pathogens to invade and induce disease in the host. Simultaneously, heat shock genes are induced, and their protein products (HSPs) are responsible for the adaptation to the higher temperature and to the hostile conditions present in host cells such as macrophages. The first heat shock gene (Hsp70) cloned in a fungal pathogen has been identified in *H. capsulatum* [25]. The response to heat shock was analyzed in two strains of *H. capsulatum* which differed considerably in thermotolerance and pathogenicity. The gene coding for a *Histoplasma* Hsp70 gene was isolated from the temperature-sensitive Downs strain (low level of virulence in mice). Using the cloned gene as a probe, transcription of Hsp70 gene at 25°C and in response to temperature shift to 34°, 37° and 40°C, temperatures that trigger the mycelial to yeast phase transition in this fungus was determined. The gene is constitutively transcribed at low levels, both in the yeast and the mycelial stages. Synthesis of Hsp70 mRNA was transiently increased 1 to 3 h after the temperature shifts. By Northern analysis, peak levels of transcription were shown to occur at 34°C in the non virulent Downs strain and at 37°C in the pathogenic G222B strain. These results are consistent with reports in which it has been shown that heat shock gene expression is part of temperature adaptation and of the associated morphological processes. The low levels of transcription of the Hsp70 gene in the Downs strain at 37°C correlated with its greater temperature sensitivity and low level of virulence [30]. Moreover, in *H. capsulatum* the spliceosome responsible for mRNA maturation is functional during infection as demonstrated by Minchiotti *et al.* [26]. These authors have isolated and characterized a heat-inducible gene, Hsp82, which contains three exons and two introns of 122 and 86 nucleotides, respectively. Contrary to what reported in *Drosophila melanogaster*, *Saccharomyces cerevisiae*, and other organisms, Hsp82 gene is properly spliced during severe heat conditions in both the temperature-sensitive Downs strain and in the temperature-tolerant G222B *Histoplasma* strains. The lack of a block in splicing is likely a general phenomenon in *Histoplasma* since the intron containing α-tubulin gene is also properly spliced at the upper temperature range [27]. The difference in mRNA splicing of the *H. capsulatum* Hsp82 and α-tubulin genes during severe heat shock conditions compared with that of intron-containing genes of other eukaryotic cells can be explained by at least two considerations. First, in *Drosophila* cells, experiments were performed under non physiological conditions, *i.e.*, exposure to a sudden change to a very high temperature followed by a rapid shift to a low temperature in order to ensure survival of the organism. Since conductivity of heat from air to animal tissue is slow, it is very unlikely that any organisms experience such dramatic temperature change in nature. One would anticipate that in these organisms the response would be gradual, *e.g.*, in plants exposed to sun, febrile response in mammals, or increase of water temperature for marine organisms. Under these circumstances, as the temperature rises, there is a steady and persistent induction of heat shock proteins. Thus, thermotolerance plays a central physiological role in the maintenance of functional cellular properties, including spliceosome activity in eukaryotic cells, and there would be no need for these cells to have a functional temperature-resistant spliceosome machinery at elevated temperatures. Pathogens such as *H.capsulatum*, on the other hand, are organisms that face a sudden and drastic environmental temperature change, *e.g.*, from one that is poikilothermic to one that is homeothermic as part of their life cycles. These organisms must adapt to dramatic shifts in temperatures and environmental stresses to invade and survive in mammalian tissues. In these organisms, a large number of genes are induced during the induction of the virulent phase, and it is reasonable that several intron-containing DNA sequences play a vital role in adaptation to the new environment. Therefore, in dimorphic organisms such as *H.capsulatum*, during heat shock response spliceosome must remain functional at the onset of infection to allow the fungus to undergo morphogenesis and survive. Regulation of transcription of heat shock genes varies among species. Stress-induced transcription requires activation of a heat shock factor (HSF), [28, 29] that binds to the heat shock promoter element (HSE), the pentanucleotide motif 5'-nGAAn-3' [30]. At least two HSEs are required for efficient transcription of the heat shock genes [31]. According to the conventional model of heat shock gene transcription, an increase in the amount of denatured proteins within cells during stress is the primary signal that induces heat shock gene transcription. Activation of HSF occurs *via* the titration of HSPs by the stress-induced formation of unfolded and denatured proteins. HSF is translocated into the nucleus where it trimerizes before activation and binds to heat shock gene promoters prior to undergoing phosphorylation [35]. However, this model does not consider that cold blooded animals and plants (that constitute more than 95% of all extant species on Earth) and microrganisms do not induce HSPs at a genetically determined temperature, but it varies when their physiological temperature fluctuates during seasonal variations or different temperature regimes [32]. Furthermore, it does not explain how very moderate temperature increments, that do not cause protein damage *in vivo*, lead to the strong induction of HSPs and to the establishment of thermotolerance [33]. Furthermore, it is well known that HSPs are present at abnormal levels in several human chronic degenerative diseases [34, 35] and their synthesis decreases with aging, while there is no evidence for modification of the kinetics or of the accumulation of denatured proteins that may explain the changes observed in heat shock gene transcription. An alternative validated model has shown that the cellular temperature-sensing mechanism is not necessarily related to the denaturation of proteins during heat stress but it is associated with the existing lipid composition and physical state (MPS) of a cell membrane [36-38]. It has been shown that MPs control temperature response in all systems analyzed, from bacteria to yeasts to higher eukaryotic cells. The response to a cold shock is mediated by a rigidification of membrane that is coupled with enhanced transcription of specific genes (including desaturases) aimed also to membrane retailoring [41]. Data from several laboratories have strongly demonstrated that membranes not only are target of stress, but also that act as sensors in activating a stress proteins [39]. MPS, under a physiological temperature regime, is regulated by environmental temperature or by diet (in mammals and birds) and, by changes in lipid unsaturation, protein-lipid ratio, composition of lipid molecular species, raft composition, *etc.*, and determines the temperature at which heat shock genes are transcribed [40, 41]. It has been shown that minor changes in lipid composition and physical state of membrane lipids trigger a dedicated stress signaling response *via* specific chemical interactions of boundary lipids with membrane proteins [43].

The proper balance between SFA and UFA present in membranes and that is the main factor in determining a given MPS [42] is determined by desaturase enzymes. Desaturases are enzymes that introduce a double bond in specific positions (Δ^6, Δ^9, Δ^{12} *etc.*) of acyl chains of long-chain saturated fatty acids (SFAs) to form unsaturated fatty acids (UFAs) in a specific position (*e.g.* in C-12 of oleic acid by Δ^{12}-desaturase), and are conserved across kingdoms. The degree of unsaturation of fatty acids affects physical-chemical properties of membrane phospholipids and associated proteins [43]. Plasma membrane micro-domains (containing special lipids) are essential for efficient signal transduction [44]. The specificity of stress genes expression can be obtained because of the particular occurrence and distribution of membrane micro-domains (rafts, claveolae, lipid shells, *etc.*) that precisely sense biological and physical signals and different forms of stress which, in turn, affect protein function [43]. Thus, simple changes in the physical properties of membrane lipids determine a physical reorganization of lipid and protein membrane components that is followed by a specific gene response aimed to compensate variations in MPS. It has been shown that a crosstalk between changes in MPS and regulation of gene expression exists, particularly for heat shock genes. Thus, an abrupt change of the MPS under stress condition redetermines the threshold temperature at which HSPs are normally synthesized. Further, members of the group of small HSPs (sHSPs) have also been shown to antagonize heat-induced membrane hyperfluidization and to stabilize the bilayer state under stress conditions [38, 45]. In *H. capsulatum* addition of palmitic (a saturated fatty acid, SFA) caused a strong induction of Hsp70 and Hsp82 mRNAs when cells were heat shocked, while treatment with oleic acid (an unsaturated fatty acid, UFA) or treatment with benzyl alcohol (BA), indused or eliminated *HSP70* and *HSP82* mRNA transcription at physiological temperatures [42].

It has also been shown that virulent strains of *H. capsulatum* are unable to induce normal amounts of HSPs when they are genetically modified, and loose the ability to survive in murine macrophages. The attenuated strain do not cause disease in a mouse model (see later section "vaccine"). The highly virulent G217B and HcD3 *H. capsulatum* strains (the latter is a G217B mutant strain carrying an extra copy of its own Δ^9-desaturase gene under the transcriptional control of the up-regulated Downs Δ^9-desaturase promoter) were grown at 34°C and 37°C. The yeast forms of the highly virulent *H. capsulatum* G217B and of HcD3 strains were grown at 34°C and 37°C to mid-log phase and then heat shocked for 30 min at different temperatures. Hsp70 and Hsp82 genes were highly expressed in *H. capsulatum* G217B (from 34° to 37°C, from 34° to 42°C, and from 37° to 42°C), whereas in strain HcD3 Hsp70 transcription was detectable only from 34° to 42°C [46]. Previously, it was shown that in *Ole1⁻*, a *S.cerevisiae* lipid mutant defective in Δ^9-desaturase activity (a mutant that does not synthesize oleic acid) when complemented with its own Δ^9-desaturase sequence, the level of Hsp70 and Hsp82 gene transcription depended on the activity of promoters cloned from different strains of *H. capsulatum* used to drive Δ^9-desaturase transcription [42]. The complemented strains, depending on the *Ole1⁻* promoter used, had different membrane fluid state, that matched with the level of transcription of heat shock genes. Furthermore, it has been shown that transformation with a heterologous Δ^{12}-desaturase or with a homologous Δ^9-desaturase gene, respectively in *Salmonella typhimurium* LT2 and in the highly virulent G217B strain of the fungus *H. capsulatum* caused major changes in membrane dynamic of these organisms. These pathogens were strongly impaired in the synthesis of major stress proteins under heat shock. These data supported the hypothesis that the perception of temperature in prokaryotic and eukaryotic pathogens is strictly controlled by membrane order and by a specific membrane lipid/protein ratio that ultimately cause transcriptional activation of heat shock genes. These results show a previously unrecognized general mode of sensing temperature variation of pathogens at the onset of infection [51]. Lack of HSP accumulation have profound consequences in virulence potential of pathogens. Genes involved in the stress response, together with virulence genes (species specific), are transcriptional activated in all pathogens at the onset of infection [47, 48]. Those gene products are directly involved in the mechanisms of invasion/adaptation, operate in a coordinate fashion, and are responsible of the capacity of a pathogen to invade, replicate and induce disease in the host [29]. This vast, coordinated and generalized genetic response allows intracellular pathogens such as *H. capsulatum* to induce the disease avoiding the immune response of the host. During a heat shock *H. capsulatum* temperature sensitive strains isolated from AIDS patients have an altered membrane lipid/protein ratio, show a sharp impairment of transcription of major heat shock genes and failure of the pathogen to survive within murine macrophages. Similarly, a genetically modified *H. capsulatum* strain with a reduced stress response (over-expressing Δ^9-desaturase), fails to survive within murine macrophages and to cause disease in a mouse model of infection [51].

Virulence Determinants of *H. capsulatum*

The ability of *H. capsulatum* to transform from a multicellular mycelium to a unicellular yeast simply by shifting the temperature from 25° to 37°C has allowed the identification of genes that are differentially expressed in response to temperature. One of the genes whose expression is higher at 37°C compared to room temperature is *RYP1* (required for yeast phase growth). Disruption of *RYP1* results in constitutive filamentous growth independent of temperature. *RYP1* protein interacts with its own promoter, suggesting the existence of a positive-feedback loop that controls the expression level of *RYP1* at 37°C. *RYP1* is a DNA-associated transcriptional regulator that in response to temperature, activates the yeast-phase-specific transcriptional program (either directly or indirectly), and inhibits the expression of mycelial phase-regulated genes. However a smaller subset of mycelial-specific genes and a small fraction of genes differentially expressed at 37°C are independent of *RYP1* expression [49]. These genes that are induced by temperature irrespective of morphology, include two homologs of peptidyl prolyl cis-trans isomerase genes [50], a homolog of the endoplasmic reticulum chaperone BiP [51], and of γ-glutamyltranspeptidase, which is thought to affect stress responses and the redox state in the endoplasmic reticulum [52]. Homologs of heat shock proteins Hsp10, Hsp30, Hsp60, and Hsp90, which play a role as chaperones at high temperature, also fall into this category. Although *Ryp1* protein does not contain defined biochemical motifs, *Ryp1* protein shows significant N-terminal homology to the *Candida albicans* Wor1 protein, a master transcriptional regulator that is required for the white-opaque phases switching [53]. *Candida* Wor1 protein associates with its own promoter as part of a feedback loop that triggers a heritable opaque state. Similar to Wor1, *Ryp1* is supposed be a transcriptional regulator that controls the switch between the *H. capsulatum* saprophytic mycelial phase to the yeast parasitic phase in response to temperature.

YPS3 is a yeast-phase-specific gene originally identified by differential hybridization technique [54]. Transcription of *YPS3* gene is not detectable in mycelia, and when the phase transition mycelium-to-yeast was induced by the temperature shift 25° to 37 °C, expression of *YPS3* was detectable as early as 24 h but not within 2 h following the temperature shift. Beside its phase specificity, *YPS3* expression varies among *H. capsulatum* strains that differ in thermotolerance and virulence. *YPS3* is expressed at high level in the virulent G217B strain and it has been suggested that it may be responsible for the rapid mycelium-to-yeast transition in laboratory conditions (approximately 9 days) [55]. In contrast, the gene is transcriptionally silent in the temperature-sensitive, attenuated Downs strain (which completes this transition in approximately 2 weeks) in both late phases and throughout transitions performed at 37° or 34°C [61]. The rapidity of the phase transitions may be an important virulence determinant *in vivo*, since only the yeast form of the organism survives in tissues [5] or within macrophages where the organism proliferates intracellularly. *YPS3* encodes a protein that is secreted in the culture medium and binds to chitin in the cell wall. Moreover, *Yps3*p released by G217B strain binds, within 5 min of exposure, to the surface of strains that do not naturally express the protein [61].

Histoplasma, like other pathogens, modifies the environment and acquire nutrients such as iron, that can promote virulence. Upon ingestion by macrophages, yeasts induce new proteins to modify rapidly the phagosomal environment, and to obtain nutrients, particularly iron, for intracellular survival [56]. Yeasts acquire Fe^{+3} after their dissociation from transferrin, a reaction that requires an acidic pH. Therefore, modulation of phagolysosomal pH controls the amount of available iron for yeast cells. A gene whose change of expression required for intracellular growth in human and murine macrophages as well for iron homeostasis and mold transition is *VMA1* (encoding for the V-ATPase catalytic subunit A) [57]. This vacuolar ATPase is responsible for generating and maintaining the acid pH in phagosome [58]. V-ATPase is a proton pump, in the phagosomal membrane, that ATP-drives translocation of protons from the cytosol into the intraphagosomal space. The elevated proton concentration is lethal for certain microorganisms. In addition, low intraphagosomal pH promotes the spontaneous dismutation of superoxide to hydrogen peroxide [59], provides optimal conditions for the activity of certain hydrolytic enzymes, and appears to be prerequisite for the process of phagosome-lysosome fusion [64]. A *vma1* mutant grew normally on iron replete medium, but not on iron deficient media, thus expression of *vma1* is required for iron homeostasis and other metabolic processes. This may affect the ability of the yeasts to produce element(s) that are responsible for inhibiting PL-fusion, or, alternatively, PL-fusion may occur after the yeasts die or become metabolically inactive. The *vma1* mutant is avirulent in a murine model of pulmonary histoplasmosis [63]. *H. capsulatum* probably has a pH sensor that monitors the pH of its intraphagosomal environment and allows the yeast to respond appropriately to changes in pH.

Histoplasma yeasts are capable of growing in a calcium-deprived environment and release large quantities of a 7.8 kDa calcium-binding protein (Cbp), the predominant protein secreted by yeast-phase cells. Cbp is an important protein to survive in a calcium-limiting environment *in vivo* (such as phagolysosomes). On the contrary, mycelia do not secrete Cbp and require calcium for growth [60]. Deletion of *Cbp1* gene greatly impairs the intracellular growth of *Histoplasma* yeasts and attenuates the ability to colonize the lung as well the virulence *in vitro* and growth in limiting calcium conditions. Purified Cbp increases the association of $^{45}CaCl_2$ with yeasts after the cells have been transferred to low-calcium medium [61]. The role of Cbp appears to be specific for pathogenesis as yeasts lacking Cbp grow normally in laboratory culture. Cbp is a homodimer protein that binds two molecules of calcium with a KD around 6 nM [62]. Furthermore, Cbp contains three disulfide bonds within each monomer, is highly stable and resistant to proteolysis and to unfolding over a range of pH values enabling Cbp to survive in the hostile acidified phagolysosome compartment [68]. For a summary of the major virulence genes identified in *H. capsulatum* see Table **2**.

Table 2: *Histoplasma capsulatum* virulence genes

Gene	Function
HSP17, HSP70, HSP82	Genes expressed during the mycelium-to-yeast transition. Strains expressing low level of these genes are not virulent in humans. Though HSP70 is the major antigen present in yeast cells, antibodies anti-HSP70 do not protect from the infection.
Δ⁹-desaturase (Δ⁹-des)	It is expressed during the mycelium-to-yeast transition. The enzyme introduces a double bond in long chain fatty moiety of the saturated fatty acid palmitic acid (18:0, SFA) to produce oleic acid, an unsaturated fatty acid (18:1, UFA). Non virulent strains of *H. capsulatum* that are T-sensitive express this gene at low level. The Δ⁹-des promoter of a non-virulent and T-sensitive strain (Downs) is an altered promoter that causes deregulation of the synthesis of UFAs (increase of UFAs and of MPS). An *Ole1⁻* mutant of *S. cerevisiae* when complemented with its own wild type gene under the control of the Δ⁹-des Downs shows an altered pattern of expression of HSPs.
Acidic lysosomal hydrolases	The coded proteins determine an increase to pH 6.5 in the phagolysosomes of macrophages.
VAM1	Encodes for the V-ATPase catalytic subunit A and is responsible for generating and maintaining the acidic pH in phagosomes.
Cbp	A *Calcium-binding protein* expressed only in yeast phase cells that allows macrophage lysis after phagocytosis.
yps, yeast phase specific genes	
yps3	Codes for a cell wall protein that has immunogenic properties.
RYP1	Mutations in this gene block *H. capsulatum* in the mycelial phase even at temperatures higher than 30°C.

Vaccine

The traditional methods to fight pathogens are not generally very effective since they are based on the use of vaccines that stimulate an immune response against attenuated strains or against one or few purified antigenic microbial proteins. The first vaccines based on purified antigens were developed for pathogens with little antigenic variability that resulted in life immunity to further exposure to the same pathogen. Eventually, problems emerged in trying to develop vaccines against pathogens with a high level of antigenic variability. In general, this results in the low efficacy of these vaccines since the natural mechanisms of protection against infective agents consist of a complex response against the entire microorganism and combined antigens. Further, vaccines made of infective attenuated microrganisms, obtained by a thermal or chemical treatment, may contain denatured antigens that do not induce the appropriate immunological response. Genetic engineering and biotechnology are making available new tools in vaccine production, allowing the molecular analysis of the immune system and its response to infectious agents. Genetic manipulation of pathogenic strains is undoubtedly safer than the classic technology based on chemical, physical or biological attenuation, because the introduction of more than a single targeted genomic mutation or deletion makes recombination events very unlikely.

As mentioned earlier HSPs have been utilized to immunize animal model of infections. Induction of HSPs is indeed a critical step in the process of adaptation of all pathogens and these proteins are the major antigens during infection. Therefore, this phenomenon has been utilized to produce vaccines based on antibodies against members of HSPs, though with limited success. Recombinant HSP70 (rHSP70) of yeast-phase *Histoplasma* did not confer protection, although has antigenic properties. A rHSP70 from *H. capsulatum* was utilized to test whether it could stimulate a cell-mediated immune response to this pathogen. Splenocytes from mice immunized with the recombinant antigen multiplied *in vitro* in response to rhsp70, in contrast to splenocytes from animals immunized with viable yeast cells that did not. Although rhsp70 is antigenic it does not protect mice for further exposure to the same virulent strain. It has been shown that rhsp70 amplifies the immune response, since yeast-immunized mice reacted *in vivo* to the antigen but cells from these animals did not recognize it [63]. In contrast, rhsp60 from *H. capsulatum* stimulated cellular immune responses and protected mice against a lethal intranasal challenge [64].

Immunization with recombinant HSP60 (rHSP60) or a region of the protein designated fragment 3 (F3) confers protection in mice against a lethal intranasal inoculation with yeast cells, and its immunological activity is similar to that of the native protein [70, 65]. Interestingly, HSP60 inhibits the binding of yeasts to macrophages in a concentration-dependent manner, but it does not promote binding to DC that recognize *Histoplasma via* the fibronectin receptor (very late antigen-5). The utilization of different receptor-ligand combinations by these cells may be explained in part by the ability of DC to inhibit the intracellular growth of yeasts, killing of the pathogen and subsequently process appropriate antigens to facilitate the induction of cell-mediated immunity [72]. On the contrary, while *Histoplasma* disrupt the normal hostile environment present within macrophages and multiplies [73], it is inhibited or killed when the cell-mediated immune defense (such as DC, lymphocytes, *etc.*) is activated.

Porta *et al.* [52] used a complete different approach to attenuate virulence of *Histoplasma* and *Salmonella*. This method has a major advantage over other techniques of attenuation since is not based on the over-expression of a specific antigen (such as HSP), it does not require the identification of species-specific genes or development of a particular genetic procedure for each pathogen. This method is, in contrast, based on a phenomenon shared by all organisms, the control of the heat shock response *via* modification of the MPS, that has extensively shown to be analogous in both prokaryotic and eukaryotic cells. The strong reduction of all members of HSPs at the onset of infection severely affect the capacity of a pathogen to adapt and thus to colonize and eventually cause disease.

As described earlier, membrane (of bacteria, yeasts, mammalian and plant cells) sense temperature variation by an abrupt change in MPS that alters the lipid/protein interactions that, in turn, are responsible for the decrease of transcriptional activation of heat shock genes [39, 74, 75]. Moreover, in all systems analyzed, the genetic (or chemical) modifications leading to an increase of MPS (more fluid membrane) reset the temperature at which heat shock transcription occurs, generally several degrees lower than in normal cells. Therefore, these cells tend to induce HSPs at normal temperature rather than at heat shock temperature. Whereas, by reducing MPS (reduced fluidity, more rigid membrane) the temperature at which heat shock gene transcription occurs is shifted to higher temperature. It has also been shown that more virulent strains of *Histoplasma* are more temperature resistant and induce heat shock response to higher temperatures, while less virulent or non pathogenic isolates have a much higher temperature sensitivity and lower heat shock response [70, 71]. Thus, a change in MPS hampers the capacity to induce the overall heat shock response, not a single HSP. This is a general response and it is not restricted to specific cells or particular conditions. The possibility to manipulate genetically the membrane physical state (which, in turn, modifies the normal temperature threshold of heat shock response) in pathogens allowed Porta *et al.* [52, 75] to manipulate the HSPs response to obtain a *H. capsulatum* strain (and *S. typhimurium*) less capable to grow inside macrophages and to induce virulence. Expression of a single gene (Δ^9-desaturase in the case of *H.capsulatum*, or desA gene or just a portion of a trans-membrane region, in *Salmonella*) caused a cascade effect. The expression of a exogenous protein in membrane pathogens changes several biochemical/biophysical properties (MPS, membrane leakage, heat capacity, *etc*). In *Histoplasma* HSP70 and HSP17 (respectively dnaK and ibpB in *Salmonella*) were used to monitor changes in HSP accumulation and showed that in both pathogens the heat shock response was severely altered. In the case of *H.capsulatum*, the attenuated strain did not cause disease in a mouse model of infection. Furthermore, when this modified strain was injected in mice induced protection if the mice were challenged with a subsequent infection with the virulent strain, suggesting that the overall antigen profile of *H. capsulatum* was unaffected or only slightly affected, thus presenting a full antigenic repertoire to the host immune system, while the capacity to express virulence genes might have been severely hampered under stress conditions.

These data in an eukaryotic pathogen (as well as in a prokaryotic pathogen) represent a novel approach to understand the relationship between adaptation and expression of virulence traits that have important applications for the development of an innovative class of new vaccines. [52, 75].

ACKNOWLEDGEMENTS

This work was supported in part by grants from MIUR/2010, Italy.

REFERENCES

[1]   Knox KS and Hage CA. Histoplasmosis. Proc Am Thorac Soc 2010; 7: 169-72.

[2]   Newman SL. Cell-mediated immunity to *Histoplasma capsulatum*. Semin Respir Infect 2001; 16: 102-8.

[3]   Gomez FJ, Pilcher-Roberts R, Alborzi A, *et al. Histoplasma capsulatum* cyclophilin a mediates attachment to dendritic cell vla-5. J Immunol 2008; 181: 7106-14.

[4]   Maresca B, Medoff J, Schlessinger D, *et al.* Regulation of dimorphism in the pathogenic fungus *Histoplasma capsulatum*. Nature 1977; 266: 447-8.

[5]   Medoff G, Kobayashi GS, Painter A, *et al.* Morphogenesis and pathogenicity of *Histoplasma capsulatum*. Infect Immun 1987; 55: 1355-8.

[6]   Kanetsuna F, Carbonell LM, Gil F, Azuma I. Chemical and ultrastructural studies on the cell walls of the yeast like and mycelial forms of *Histoplasma capsulatum*. Mycopath Mycol Appl 1974; 54: 1-13.

[7]   Palma AS, Feizi T, Zhang Y, *et al.* Ligands for the beta-glucan receptor, Dectin-1, assigned using "designer" microarrays of oligosaccharide probes (neoglycolipids) generated from glucan polysaccharides. J Biol Chem 2006; 281: 5771-9.

[8]   Gorocica P, Taylor ML, Alvarado-Vásquez N, *et al.* The interaction between *Histoplasma capsulatum* cell wall carbohydrates and host components: relevance in the immunomodulatory role of histoplasmosis. Mem Inst Oswaldo Cruz, Rio de Janeiro 2009; 104: 492-6.

[9]   Smits GJ, van den Ende H, Klis FM. Differential regulation of cell wall biogenesis during growth and development in yeast. Microbiology 2001; 147: 781-94.

[10]   Davis TE Jr, Domer JE, Li YT. Cell wall studies of *Histoplasma capsulatum* and *Blastomyces dermatitidis* using autologous and heterologous enzymes. Infect Immun 1977; 15: 978-87.

[11]   Rappleye CA, Engle JT, Goldman WE. RNA interference in *Histoplasma capsulatum* demonstrates a role for alpha-(1,3)-glucan in virulence. Mol Microbiol 2004; 53: 153-65.

[12]   Edwards JA, Alore EA, Rappleye CA. The yeast-phase virulence requirement for α-glucan synthase differs among *Histoplasma capsulatum* chemotypes. Eukaryot Cell 2011; 10: 87-97.

[13]   Marion CL, Rappleye CA, Engle JT, *et al.* An alpha-(1,4)-amylase is essential for alpha-(1,3)-glucan production and virulence in *Histoplasma capsulatum*. Mol Microbiol 2006; 62: 970-83.

[14]   Klimpel KR, Goldman WE. Cell Walls from Avirulent Variants of *Histoplasma capsulatum* Lack α-(1,3)-Glucan. Infect Immun 1988; 56: 2997-3000.

[15]   Goodridge HS, Wolf AJ, Underhill DM. β-glucan recognition by the innate immune system Immunological Reviews 2009; 230: 38-50.

[16]   Willment JA, Gordon S, Brown GD. Characterization of the human beta-glucan receptor and its alternatively spliced isoforms. J Biol Chem 2001; 276: 43818-23.

[17]   Rappleye CA, Groppe Eissenberg L, Goldman WE. *Histoplasma capsulatum* α-(1,3)-glucan blocks innate immune recognition by the β-glucan receptor. Proc Natl Acad Sci USA. 2007; 104: 1366-70.

[18]   Newman SL, Bucher C, Rhodes J, *et al.* Phagocytosis of *Histoplasma capsulatum* yeasts and microconidia by human cultured macrophages and alveolar macrophages. Cellular cytoskeleton requirement for attachment and ingestion. J Clin Invest 1990; 85: 223-30.

[19]   Bullock WE, Wright SD. Role of the adherence-promoting receptors, CR3, LFA-1, and p150,95, in binding of *Histoplasma capsulatum* by human macrophages. J Exp Med 1987; 165: 195-10.

[20]   Long KH, Gomez FJ, Morris RE, *et al.* Identification of heat shock protein 60 as the ligand on *Histoplasma capsulatum* that mediates binding to CD18 receptors on human macrophages. J Immunol 2003; 170: 487-94.

[21]   Kubota H, Hynes G, Willison K. The chaperonin containing t-complex polypeptide 1 (TCP-1): multisubunit machinery assisting in protein folding and assembly in the eukaryotic cytosol. Eur J Biochem 1995; 230: 3-16.

[22]   Deepe GS Jr, Gibbons RS.. Cellular and molecular regulation of vaccination with heat shock protein 60 from *Histoplasma capsulatum*. Infect Immun 2002; 70: 3759-67.

[23]   Eissenberg LG, Schlesinger PH, Goldman WE. Phagosome-lysosome fusion in P388D1 macrophages infected with *Histoplasma capsulatum*. J Leukoc Biol 1988; 43: 483-91.

[24]   Eissenberg LG, Goldman WE, Schlesinger PH. *Histoplasma capsulatum* modulates the acidification of phagolysosomes. J Exp Med 1993; 177: 1605-11.

[25]   Gildea LA, Ciraolo GM, Morris RE, Newman SL. Human dendritic cell activity against *Histoplasma capsulatum* is mediated *via* phagolysosomal fusion. Infect Immun 2005; 73: 6803-11.

[26] Newman SL, Gootee L, Hilty J, *et al.* Human macrophages do not require phagosome acidification to mediate fungistatic/fungicidal activity against *Histoplasma capsulatum.* J Immunol 2006; 176: 1806-13.

[27] Lindquist S. The heat shock response. Annu Rev Biochem 1986; 5: 1151-91.

[28] Ellis J. Proteins as molecular chaperones. Nature 1987; 328: 378-9.

[29] Vigh L, Maresca B and Harvood JL. Does the membrane's physical state control the expression of heat shock and other genes? Trends Biochem Sci 1998; 23: 369-74.

[30] Groisman EA and Ochman H. How *Salmonella* became a pathogen.Trends Microbiol 1997; 5: 343-9.

[31] Caruso M, Sacco M, Medoff G, *et al.* Heat shock 70 gene is differentially expressed in *Histoplasma capsulatum* strains with different levels of thermotolerance and pathogenicity. Mol Microbiol 1987; 1: 151-8.

[32] Minchiotti G, Gargano S, Maresca B.The intron-containing hsp82 gene of the dimorphic pathogenic fungus *Histoplasma capsulatum* is properly spliced in severe heat shock conditions. Mol Cell Biol 1991; 11: 5624-30.

[33] Minchiotti G, Gargano S, Maresca B. Molecular cloning and expression of hsp82 gene of the dimorphic pathogenic fungus *Histoplasma capsulatum.* Biochim Biophys Acta 1992; 1131: 103-7.

[34] Morimoto RI, Sarge KD, Abravaya K. Transcriptional regulation of heat shock genes. J Biol Chem 1992; 267: 21987-90.

[35] Voellmy R. Sensing stress and responding to stress. EXS. 1996; 77: 121-37.

[36] Morimoto RI. Regulation of the heat shock transcriptional response: cross talk between a family of heat shock factors, molecular chaperones, and negative regulators. Genes Dev 1998; 12: 3788-96.

[37] Sorger PK, Lewis MJ and Pelham HR. Heat shock factor is regulated differently in yeast and HeLa cells. Nature 1987; 329: 81-4.

[38] Tomanek L and Somero GN. Evolutionary and acclimation-induced variation in the heat-shock responses of congeneric marine snails (genus Tegula) from different thermal habitats: implications for limits of thermotolerance and biogeography. J Exp Biol 1999; 202: 2925-36.

[39] Saidi Y, Finka A, Muriset M, *et al.* The heat shock response in moss plants is regulated by specific calcium-permeable channels in the plasma membrane. Plant Cell 2009; 21: 2829-43.

[40] Calderwood SK, Khaleque MA, Sawyer DB, *et al.* Heat shock proteins in cancer: chaperones of tumorigenesis. Trends Biochem Sc 2006; 31: 164-72.

[41] Delogu G, Signore M, Mechelli A, *et al.* Heat shock proteins and their role in heart injury. Curr Opin Crit Care 2002; 8: 411-6.

[42] Balogh G, Horváth I, Nagy E, *et al.* The hyperfluidization of mammalian cell membranes acts as a signal to initiate the heat shock protein response. FEBS J 2005; 272: 6077-86.

[43] Carratù L, Franceschelli S, Pardini CL, *et al.* Membrane lipid perturbation modifies the set point of heat shock response in yeast. Proc Nat Acad Sci USA 1996; 93: 3870-5.

[44] Vigh L, Escriba P, Sonnleitner A, *et al.* The significance of lipid composition for membrane activity: new concepts and ways of assessing function. Prog Lipid Res 2005; 44: 303-44.

[45] Hofmann NR. The PlasmaMembrane as First Responder to Heat Stress.The Plant Cell 2009; 21: 2544.

[46] Szalontai B, Nishiyama Y, Gombos Z, *et al.* Membrane dynamics as seen by Fourier transform infrared spectroscopy in a cyanobacterium, Synechocystis PCC 6803. The effects of lipid unsaturation and the protein-to-lipid ratio. Biochim Biophys Acta 2000; 1509: 409-19.

[47] Vigh L, Maresca B and Harvood JL. Does the membrane's physical state control the expression of heat shock and other genes? Trends Biochem Sci 1998; 23: 369-74.

[48] Cossins AR. Homeoviscous adaptation of biological membranes and its functional significance, In: Cossins AR ed., Temperature adaptation of biological membranes. London, Portland Press, UK, 1994; pp. 63-76.

[49] Vigh L, Literati NP, Horvath I, *et al.* Bimoclomol: a nontoxic, hydroxylamine derivative with stress protein-inducing activity and cytoprotective effects. Nature Med 1997; 3: 1150-4.

[50] Shaw AS. Lipid rafts: now you see them, now you don't. Nature Immunol 2006; 7: 1139-42.

[51] Tsvetkova NM, Horváth I, Török Z, *et al.* Small heat-shock proteins regulate membrane lipid polymorphism. Proc Natl Acad Sci USA 2002; 99: 13504-9.

[52] Porta A, Eletto A, Török Z, *et al.* Changes in membrane fluid state and heat shock response cause attenuation of virulence. J Bacteriol 2010; 192: 1999-05.

[53] Acharya P, Kumar R, Tatu U. Chaperoning a cellular upheaval in malaria: heat shock proteins in *Plasmodium falciparum.* Mol Biochem Parasitol 2007; 153: 85-94.

[54] Henderson B, Allan E, Coates AR. Stress wars: the direct role of host and bacterial molecular chaperones in bacterial infection. Infect Immun 2006; 74: 3693-06.

[55]   Nguyen VQ, Sil A. Temperature-induced switch to the pathogenic yeast form of *Histoplasma capsulatum* requires *Rypl*, a conserved transcriptional regulator. Proc Natl Acad Sci USA 2008; 105: 4880-5.

[56]   Nagradova N. Enzymes catalyzing protein folding and their cellular functions. Curr Protein Pept Sci 2007; 8: 273-82.

[57]   Kleizen B, Braakman I. Protein folding and quality control in the endoplasmic reticulum. Curr Opin Cell Biol. 2004; 16: 343-9.

[58]   Kinlough CL, Poland PA, Bruns JB, *et al.* Gamma-glutamyltranspeptidase: disulfide bridges, propeptide cleavage, and activation in the endoplasmic reticulum. Methods Enzymol 2005; 401: 426-49.

[59]   Huang G, Wang H, Chou S, *et al.* Bistable expression of WOR1, a master regulator of white-opaque switching in Candida albicans . Proc Natl Acad Sci USA 2006; 103: 12813-8.

[60]   Keath EJ, Painter AA, Kobayashi GS, *et al.* Variable expression of a yeast-phase-specific gene in *Histoplasma capsulatum* strains differing in thermotolerance and virulence. Infect Immun 1989; 57: 1384-90.

[61]   Bohse ML, Woods JP. Surface localization of the *Yps3*p protein of *Histoplasma capsulatum*. Eukaryot Cell 2005; 4: 685-93.

[62]   Newman SL, Gootee L, Brunner G, *et al.* Chloroquine induces human macrophage killing of *Histoplasma capsulatum* by limiting the availability of intracellular iron and is therapeutic in a murine model of histoplasmosis. J Clin Invest 1994; 93: 1422-29.

[63]   Hilty J, Smulian AG, Newman SL. The *Histoplasma capsulatum* vacuolar ATPase is required for iron homeostasis, intracellular replication in macrophages and virulence in a murine model of histoplasmosis. Mol Microbiol 2008; 70: 127-39.

[64]   Lukacs GL, Rotstein OD, Grinstein S. Determinants of phagolysosomal pH in macrophages. J Biol Chem 1991; 266: 24540-48.

[65]   Fridovich I. The biology of oxygen radicals. Science 1978; 201: 875-80.

[66]   Batanghari JW, Goldman WE. Calcium dependence and binding in cultures of *Histoplasma capsulatum*. Infect Immun 1997; 65: 5257-61.

[67]   Sebghati TS, Engle JT, Goldman WE. Intracellular Parasitism by *Histoplasma capsulatum*: Fungal Virulence and Calcium Dependence. Science 2000; 290: 1368-72.

[68]   Beck MR, Dekoster GT, Hambly DM, *et al.* Structural features responsible for the biological stability of Histoplasma's virulence factor Cbp, Biochemistry 2008; 47: 4427-38.

[69]   Allendoerfer R, Maresca B, Deepe GS Jr. Cellular immune responses to recombinant heat shock protein 70 from *Histoplasma* capsulatum. Infect Immun 1996; 64: 4123-8.

[70]   Gomez FJ, Allendoerfer R, Deepe GS Jr. Vaccination with recombinant heat shock protein 60 from *Histoplasma capsulatum* protects mice against pulmonary histoplasmosis. Infect Immun 1995; 63: 2587-95.

[71]   Scheckelhoff M, Deepe GS Jr. The protective immune response to heat shock protein 60 of *Histoplasma capsulatum* is mediated by a subset of V beta 8.1/8.2+ T cells. J Immunol 2002; 169: 5818-26.

[72]   Gildea LA, Morris RE, Newman SL. *Histoplasma capsulatum* yeasts are phagocytosed *via* very late antigen-5, killed, and processed for antigen presentation by human dendritic cells. J Immunol 2001; 166: 1049-56.

[73]   Newman SL, Gootee L, Bucher C, *et al.* Inhibition of intracellular growth of *Histoplasma capsulatum* yeast cells by cytokine-activated human monocytes and macrophages. Infect Immun 1991; 59: 737-41.

[74]   Török Z, Tsvetkova NM, Balogh G, *et al.* Heat shock protein coinducers with no effect on protein denaturation specifically modulate the membrane lipid phase. Proc Natl Acad Sci USA. 2003; 100: 3131-6.

[75]   Porta A, Török Z, Horvath I, *et al.* Genetic modification of the *Salmonella* membrane physical state alters the pattern of heat shock response. J Bacteriol 2010; 192: 1988-98.

[76]   Medoff G., Maresca B., Lambowitz AM, *et al.* Correlation between pathogenicity and temperature sensitivity in different strains of *H. capsulatum*. J Clin Investig 1986; 78: 1638-47.

[77]   Lambowitz AM, Kobayashi GS, Painter A, *et al.* Possible relationship of morphogenesis in the pathogenic fungus, *Histoplasma capsulatum* to heat shock response. Nature 1983; 303: 806-8.

CHAPTER 5

Development and Dimorphism of the Yeast *Yarrowia lipolytica*

Juan Francisco Jiménez-Bremont[1],*, Aída Araceli Rodríguez-Hernández[1] and Margarita Rodríguez-Kessler[2]

[1]*División de Biología Molecular, Instituto Potosino de Investigación Científica y Tecnológica (IPICYT). San Luis Potosí, México and* [2]*Facultad de Ciencias, Universidad Autonóma de San Luis Potosí (UASLP), San Luis Potosí, México*

Abstract: *Yarrowia lipolytica* is one of the most important non-conventional yeasts that belongs to the Dipodascaceae family of hemiascomycetous fungi. *Y. lipolytica* is used for both academic and biotechnological applications. Respect to biotech traits, *Y. lipolytica* is widely used in production of single-cell protein, organic acids and enzymes, also it utilized as heterologous protein expression system, and bioremediation issues, among others. On the other hand, *Y. lipolytica* has become a model used to study several biological themes such as: dimorphism, protein secretion, gene manipulation, protein expression, peroxisome biogenesis, physiology, genetics, degradation of hydrophobic substrates, and lipid accumulation, etc. *Y. lipolytica* is a dimorphic organism that grows as a mixture of yeast-like and short mycelial cells. This behavior is influenced by pH, carbon and nitrogen sources, blood serum, citrate, polyamines and anaerobic stress. In the present chapter, we review many important regulators involved in the dimorphic switch in *Y. lipolytica*. All the knowledge about the yeast-to-hypha transition that has been obtained from this non-pathogenic yeast, providing information that undoubtedly will be useful for the understanding of this phenomenon in important pathogenic organisms.

Keywords: *Yarrowia lipolytica*, biotechnology, dimorphism, morphogenesis, polyamines, hyphae, yeast, pH, Pal/Rim pathway, non-conventional yeast, cell polarity, secretion.

INTRODUCTORY ASPECTS

Yarrowia lipolytica, a hemiascomycetous fungus, is one of the most important non-conventional yeasts used for both academic and biotechnological applications. This yeast presents the ability to extensively biodegrade oils and many hydrocarbons. The natural habitats of this yeast are oil-polluted environments and dairy foods such as cheese, yogurt, kefir, shoyu, meat and shrimps [1].

Y. lipolytica was originally named as *Candida lipolytica*, but with the observation that some mixtures of strains had the ability to form ascospores, *i.e.,* that it has a sexual cycle, the fungus was reclassified and named *Endomycopsis lipolytica, Saccharomycopsis lipolytica*, and finally *Yarrowia lipolytica* [2, 3]. This yeast has the ability to metabolize diverse industrial and agro-industrial residues for the production of single-cell protein, single-cell oil, organic acids such as citric and 2-ketoglutaric [4-7]. For citric acid production, *Y. lipolytica* has been classified as Generally Regarded As Safe (GRAS) by the American Food and Drug Administration (FDA).

Another important biotechnological application of *Y. lipolytica* is the production of enzymes, such as lipases, esterases, proteases, and phosphatases, since this yeast has the ability to naturally secrete proteins [8]. Also, it has been utilized in the treatment of wastewater [9], exemplified by the use of a tropical marine strain for the degradation of aliphatic hydrocarbons and triglycerides in palm oil mill effluent. Currently, *Y. lipolytica* has an important strain collection including wild-type, mutants and recombinant strains for different environmental and industrial applications [10].

Adress correspondence to Juan Francisco Jiménez-Bremont: División de Biología Molecular, Instituto Potosino de Investigación científica y Tecnológica (IPICYT), San Luis Potosí, SLP, Camino a la Presa de San José 2055, Apartado Postal 3-74 Tangamanga, C.P. 78216, San Luis Potosí, SLP, México. Email: jbremont@ipicyt.edu.mx

Y. lipolytica has peculiar characteristics that make it an excellent system for overexpression of heterologous proteins [1, 11]. Furthermore, developments in genetic engineering and molecular biology make the non conventional yeast *Y. lipolytica* as one of the most promising hosts for efficient heterologous protein expression [12, 13, 8]. This yeast has gained attraction as a heterologous expression system because of its advantages in comparison to *Saccharomyces cerevisiae*, which has presented limitations as a host for heterologous protein production [1]. Among the advantages that *Y. lipolytica* strains present are: production competitive with other systems, a remarkable regularity of performance in the efficient secretion of heterologous proteins, absence of hyperglycosylation, and normal biological activity [14].

As above indicated, *Y. lipolytica* has been used for remediation of the environment contaminated by various pollutants, including heavy metals and hydrocarbons; *e.g.* this strict aerobic yeast was used in the bioremediation of marine and soil environments contaminated by petroleum byproducts [15-18]. *Y. lipolytica* has the advantage that it can survive in extreme environments such as those containing high concentrations of NaCl, heavy metals and different pollutants [19-24]. *Y. lipolytica* cells can bind large amounts of heavy metals and produce specific protein metallothionein as protective agent [21]. It also produces brown extracellular pigments, such as melanins [25]; that is another defense mechanism to protect it against metal-induced oxidative stress. Regarding salt tolerance, it may be indicated that several strains have been isolated from hypersaline and marine locations [23, 24].

In addition to the biotech traits, *Y. lipolytica* has become a model system used to study several biological issues such as: dimorphism, protein secretion, gene manipulation, protein expression, peroxisome biogenesis, physiology, genetics, degradation of hydrophobic substrates, and lipid accumulation [1, 8, 13 26-31]. Due to its biological relevance, the entire sequence of the six chromosomes in *Y. lipolytica* has been obtained. The genome sequence with a size of 20.5 Mb, has revealed that this yeast is distantly related to the conventional yeast *S. cerevisiae* whose genome has a size of 12.4Mb divided in sixteen chromosomes [32].

Owing to all these characteristics, the study of the fungus has given rise to extensive reviews on its different characteristics: physiology, genetics, molecular biology, and biotechnical application, *etc.* [1, 8, 26, 27, 30, 31]. In the present chapter we will discuss only briefly these aspects of this non-conventional yeast, and will dedicate a deeper discussion to the physiology and biochemical mechanisms related to dimorphism, according to the general theme of the volume.

Dimorphism in Fungi

Several fungal species exhibit a morphological transition (dimorphism), expressed as the capacity to grow in two distinct morphological forms: as yeast cells or as filamentous hyphae. Dimorphism is not restricted to a special group of fungi; instead dimorphic species are found among the main taxa of the kingdom [33, 34]. This phenomenon is believed to constitute a mechanism of response to environmental conditions and represents an important attribute for the development of virulence by a number of pathogenic fungi, including *Histoplasma capsulatum*, *Candida albicans*, *Ustilago maydis*, *etc.* These fungi display different morphologies in their saprophytic and virulent phases of growth. Understanding the mechanisms that regulate the dimorphic switch will provide valuable information for the treatment of fungal infections. On the other hand, this phenomenon has been reported in nonpathogenic fungi such as *Benjaminiella poitrasii* [35], *Mucor circinelloides* [36] and *Y. lipolytica* [27].

Several conditions have been described that induce the dimorphic transition of yeast-to-mycelium or mycelium-to-yeast. Among these we may cite, changes in pH, temperature, nutritional status, the gaseous atmosphere of growth, or the presence of specific compounds in the culture media such as N-acetylglucosamine (GlcNAc), blood serum, *etc.* [29].

Dimorphism in Yarrowia lipolytica

Yarrowia lipolytica is a dimorphic organism that grows as a mixture of yeast-like and short mycelial cells in liquid and solid media [37; Fig. **1**]. Although *Y. lipolytica* is a non pathogenic fungus, this yeast has been used as an excellent model for the study of fungal dimorphism, since it may be subjected to genetic manipulation [38] and transformation [12, 39].

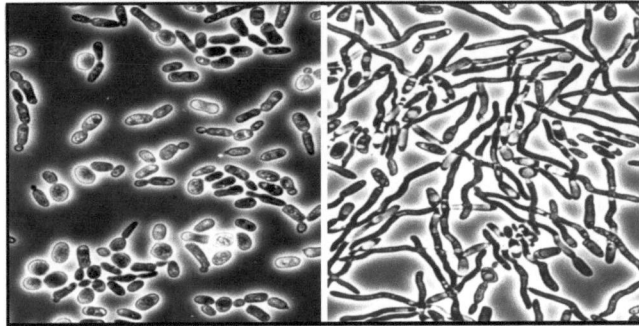

Figure 1: Morphology of *Yarrowia lipolytica* (P01A strain) grown as yeast in YNB-glucose medium (left), or as mycelium in YNB-*N*-acetylglucosamine medium (right).

Factors Involved in Dimorphism in Yarrowia lipolytica

Y. lipolytica suffers morphological changes under a variety of conditions, indicating the presence of different signaling pathways for dimorphic transition [40]. Dimorphism in *Y. lipolytica* is influenced by the interplay of pH [29], carbon and nitrogen sources [29, 41-43], blood serum [27, 44], citrate and anaerobic stress [29], all of which have been reported to regulate the dimorphic switch and the subsequent adaptation of the fungus to its natural ecological niche.

Ruiz-Herrera and Sentandreu [29] analyzed this phenomenon in detail, and reported that at pH around neutrality, *Y. lipolytica* is able to develop in the mycelial form, whereas when pH decreased until a value of 3, mycelium growth became almost null, and it grew as a population of yeast cells. *C. albicans* responds the same way as *Yarrowia*, growing in the form of mycelium at neutral pH and yeast-like in an acid medium. However, in *C. albicans* when GlcNAc was used as carbon source, mycelium formation occurred independently of the pH of the medium [45]. Interestingly, the phytopathogenic fungus *U. maydis* responds to pH in an opposite way, growing as mycelium at an acid pH (with an optimal value of pH 3), and at neutral pH a yeast-like morphology is developed [46].

In fungi the main transduction pathway involved in control by pH is the Rim/Pal pathway. Analysis of the phenotype of different strains of *Y. lipolytica* carrying mutations in genes encoding members of this pathway (*RIM9*, *RIM13*, and *RIM101*) were not affected in the dimorphic transition [47]. These results demonstrated that the Rim pathway was not involved in the regulation of dimorphism by pH in *Y. lipolytica*. A similar result was obtained with *U. maydis*, where capacity of *rim101* mutants to carry out their dimorphic transition is not affected [48].

In respect to the role of the carbon source, a study of the environmental factors involved in the regulation of the dimorphic transition indicates that use of *N*-acetylglucosamine (GlcNAc) induced mycelial growth [41]. The formation of hyphae is achieved changing glucose for GlcNAc as the carbon source, or adding serum to the culture medium [44, 40], and also when this yeast is grown in rich media. Since in the work reported by Ruiz-Herrera and Sentandreu [29], the strain used was different to that employed by Rodriguez and Domínguez [41], and glucose was as efficient as *N*-acetylglucosamine in inducing hyphal growth, it appears that the dimorphic capacity and the role of the different effectors is strain-dependent in the *Y. lipolytica* species.

Dimorphism and Polyamine Metabolism in Yarrowia lipolytica

Polyamines (PAs) are ubiquitous and essential aliphatic cations present in most prokaryotes and all eukaryotic organisms [49-51]. Different authors have demonstrated that polyamines are required for cell growth and development [49, 52-57], although their precise mode of action is unknown. These polycationic molecules with a hydrocarbon backbone and multiple amino groups play different roles in DNA protection: from enzymatic degradation, X-ray irradiation, mechanical shearing and oxidative damage [58, 59]. At physiological pH, PAs are found as protonated, positively charged molecules containing two (diamine), three (triamine) or four (tetraamine) amine groups, what favors their electrostatic interaction with several negatively-charged macromolecules such as nucleic acids, proteins and lipids [60, 61]. At the cellular level, PAs participate in diverse fundamental processes such as transcription, translation, DNA replication, chromatin condensation, cell signaling, cell division and differentiation, senescence and cell death. In addition, diverse roles in membrane stabilization, ion channel regulation, cation-anion balance, modulation of enzyme activities, and protein modification have been also described [61-63].

In fungi, polyamines are synthesized by a pathway initiated by ornithine decarboxylase (ODC) with formation of putrescine from ornithine, continuing the synthesis of spermidine by the reaction of putrescine with decarboxylated S-adenosylmethionine (Samdc), catalyzed by spermidine synthase (SPDS). In a following step, the reaction of spermidine with Samdc catalyzed by spermine synthase (SPMS) gives rise to spermine. Most fungi do not contain spermine, but only putrescine and spermidine, however *Y. lipolytica* is an exception, containing significant amounts of spermine [42].

In fungi, PAs are essential for distinct differentiation processes, such as dimorphism, spore germination and sporulation [56, 64]. It has been reported that spore germination is accompanied by an increase in ODC activity and cellular polyamine levels in diverse fungi [49, 56, 65-68].

The role of polyamine metabolism in dimorphism has been analyzed in different fungal species such as: *Mucor* spp [67, 68], *Candida albicans* [69] and *Ustilago maydis* [42, 70]. In *Y. lipolytica*, transient increases in the ODC levels and polyamines have been shown to take place during the yeast-to-hypha transition [42]. In addition, the use of the specific inhibitors of ODC, diamino butanone (DAB) and difluoro methylornithine (DFMO), block the yeast-to-mycelium dimorphic transition, not affecting significantly cell growth [42]. This effect was reversed with the application of putrescine, and also showed a time-dependent behavior, since addition of ODC inhibitors after a critical period, which coincided with a transient increase in ODC activity and polyamine concentrations, had no effect on mycelium formation [42].

Stronger evidence of the role of polyamines in the yeast-to-hypha dimorphic transition in *Y. lipolytica* was obtained through the characterization of ODC null mutants [37]. These mutants were auxotrophic for polyamines and after depletion of the polyamine pool, they grew only in the presence of added exogenous putrescine. On the other hand, Jimenez-Bremont *et al.* [37] observed that polyamine concentrations necessary to ensure vegetative growth were lower than those required to sustain the dimorphic transition. Similar results were obtained with *odc* mutants of *U. maydis* [70] and *C. albicans* [57].

To eliminate the possibility that the differential effect of the putrescine added to the medium on growth and dimorphism were due to permeability problems, experiments were devised to analyze whether polyamines synthesized into the *Y. lipolytica* cells, equally gave rise to a difference between growth and dimorphism of the *odc* mutants. To this aim, an *odc* mutant was transformed with a plasmid carrying the wild-type *ODC* gene under the control of the *Y. lipolytica* inducible promoter of MTPI-II gene (encoding metallothionein) that responds to copper addition. Transformants were able to grow in the absence of Cu^{++}, due to the basal expression of the *ODC* gene, but only in the yeast form. Cu^{++} addition to the medium was required for mycelium formation, reaching a maximum at a copper concentration of 2 mM, after which the inhibitory concentration of copper became notorious [37].

Genes Involved in Dimorphism in Yarrowia lipolytica

Several genes have been characterized with a role in dimorphism in *Yarrowia lipolytica* including *STE11, HOY1, MHY1, RAC1, TUP1, CLA4, BMH1*, among others. Cervantes-Chávez and Ruiz-Herrera [71] studied the role of the *STE11*-dependent MAPK pathway in the dimorphic transition of *Y. lipolytica*. The *Y. lipolytica Δste11* mutants grew constitutively in the yeast-like form, indicating that the pathway was somewhat involved in the transfer of the pH signal that induced the growth of the mycelial form.

Interestingly, preliminary data had demonstrated that the addition of cAMP to *Y. lipolytica* under conditions that normally induce mycelial growth, inhibited the process, and the fungus grew in the yeast-like form [29]. This result contrasted with those obtained with *C. albicans* where cAMP induces mycelial growth, and suggested that the MAPKKK and PKA pathways were antagonists. Further evidence was obtained when attempts were made to obtain mutants defective in the regulatory subunit of the *Y. lipolytica* PKA. This mutation proved to be lethal, but merodiploids carrying several copies of the wild-type gene in a self-replicating plasmid proved to be more resistant to the inhibitory effect on dimorphism of cAMP [72]. On the other hand, mutation of the gene encoding the catalytic subunit of PKA (*TPK1*) gave the final evidence that the PKA pathway was antagonist to the MAPK pathway in regards to dimorphism, since strains carrying this mutation were monomorphic, growing only in the mycelial form [73]. The observation that double mutants (Δ*ste11*/Δ*tpk1*) grew only in the yeast-like form indicated that this is the default growth pattern of the fungus.

Torres-Guzman and Dominguez [74] reported in *Y. lipolytica* the homeogene *HOY1*, which shows some regions of homology with the *S. cerevisiae* transcriptional activator Pho2p, and reported that it was differentially regulated during the yeast-to-hypha transition. The authors suggest that the *HOY1* gene was involved in the developmental pathway that controls hypha formation in a similar way to the temporal and spatial patterns that have been described for mouse limb and chicken wing development, regulated by homeogenes having a low but significant homology to *HOY1*.

The *MHY1* gene encodes a C_2H_2-type zinc finger protein, which is able to bind to a STRE sequence, and is involved in the yeast-to-hypha transition of *Y. lipolytica* [75]. During the yeast-to-hypha dimorphic transition, the expression of *MHY1* is strongly increased, and Mhy1 is concentrated in the nuclei of actively growing cells found at the hyphal tip. The disruption of *MHY1* results in complete abolition of both hyphal and pseudohyphal growth [75]. On the other hand, Hurtado [76] reported the characterization of the gene *YlRAC1*, which encodes a GTPase of the Rho family. *YlRAC1* play an important role in the regulation of hyphal growth, but is not essential for cell viability or actin organization. The expression of *YlRAC1* increases steadily during the yeast-to-hypha transition, and *ylrac1* mutants are able to form pseudohyphae but not true hyphae. In *C. albicans*, another homolog of Rac (Rac1p) was identified to be essential for filamentous growth [77].

The gene *TUP1* in *Y. lipolytica* showed to be involved in the formation of pseudohyphae [78]. The *TUP1* homologue in *C. albicans* has demonstrated to play a role in both the regulation of the bud-to-hypha transition and the transcription of a variety of genes, including genes involved in virulence [79]. Other homologue is *TUP1* in *S. cerevisae*, whose mutation affects different phenotypic traits including flocculence, sporulation and sexuality of MATα cells. These authors propose that the *TUP1* gene inhibits a class of transcriptional activator proteins [80].

Cla4 protein kinase belongs to the large family of p21-activated protein kinases (PAKs). In *Y. lipolytica*, deletion of the *CLA4* gene is not lethal, but completely eliminates the ability to form filaments and to invade agar. *cla4* mutants exhibit an aberrant distribution of chitin in the cell wall, indicating a possible role for the Cla4 protein kinase in the maintenance of cell polarity in *Y. lipolytica* [81]. In *U. maydis*, *cla4* mutants are unable to induce pathogenic development in plants and to display filamentous growth in a mating reaction, although they are still able to secrete pheromone and to undergo cell fusion with wild-type cells. It was proposed that Cla4 is involved in the regulation of cell polarity during budding and filamentation [82].

The *YlBEM1* gene is essential for the yeast-to-hypha transition in *Y. lipolytica*. The *YlBEM1* gene encodes a protein of 639 amino acids and its transcription is significantly increased during the dimorphic transition. Cells deleted in *BEM1* are viable but shows severe morphological defects. These mutants though unable to form hyphae in liquid and solid media, nevertheless form pseudohyphae to a reduced extent on agar plates [83].

Hurtado and Rachubinski [83] identified the *YlBMH1* gene that encodes a 14-3-3 protein and whose transcription levels are increased during the yeast-to-hypha transition. The 14-3-3 proteins are members of a large family of highly conserved and ubiquitously-expressed proteins, but whose exact functions have remained unclear. *Y. lipolytica* strains with mutations in 14-3-3 proteins either develop abnormal filaments or fail to differentiate completely. Also, in *C. albicans*, 14-3-3 functions have been associated with filamentous or pseudohyphal growth [84].

CONCLUSION

It is clear that dimorphism in fungi is a complex but also a very interesting phenomenon. Despite its importance, relatively little is known about this process in the dimorphic yeast *Yarrowia lipolytica*; in spite of the fact that this yeast of biotechnological interest is an excellent model for the study of fungal dimorphism, since it is amenable to the modern techniques of molecular biology and genetics. *Y. lipolytica* has also several additional positive features: it is easy to handle, it may be subjected to genetic manipulation, it has very high transformation efficiency, and a precise targeting of monocopy integration into the genome, it is easy to obtain mutants by homologous recombination, its complete genome sequence is known, *etc.* Wise use of all these techniques will allow knowing how *Y. lipolytica* cells carry out the dimorphic switch, providing information that undoubtedly will be useful for the understanding of this phenomenon in important pathogenic organisms.

REFERENCES

[1] Barth G and Gaillardin C. Physiology and genetics of the dimorphic fungus *Yarrowia lipolytica*. FEMS Microbiol Rev 1997; 19: 219-237.

[2] Yarrow D. Four new combinations in yeast. Antonie Leeuenhoek J. 1972; 38: 357-360.

[3] Van der Walt JP and Von Arx JA. The yeast genus *Yarrowia* gen nov Ant Van Leeuwenhoek 1980; 46:517-521.

[4] Papanikolaou SI, Chevalot M, Komaitis I, Marc and Aggelis G. Single cell oil production by *Yarrowia lipolytica* growing on an industrial derivative of animal fat in batch cultures. Appl Microbiol Biotechnol 2002; 58: 308-312.

[5] Finogenova TV, Morgunov IG, Kamzolova SV, Chernyavskaya OG. Organic acid production by the yeast *Yarrowia lipolytica*: a review of prospects. Appl Biochem Microbiol 2005; 41: 418-425.

[6] Rymowicz W, Rywinska A, Zarowska B. Biosynthesisofcitricacidfrom crude glycerol by *Yarrowia lipolytica* in repeated – batch cultivations. J Biotechnol 2007; 131: 149-150.

[7] Rywinska A, Rymowicz W, Zarowska B, Wojtatowicz M. Biosynthesis of citric acid from glycerol by acetate mutants of *Yarrowia lipolytica* in Fed-Batch fermentation. Food Technol Biotechnol 2009; 47: 1-6.

[8] Madzak C, Gaillardin C. and Beckerich JM. Heterologous protein expression and secretion in the non- conventional yeast *Yarrowia lipolytica*: a review. J Biotechnol 2004; 109: 63-81.

[9] Oswal N, Sarma PM, Zinjarde SS and Pant A. Palm oil mill effluent treatment by a tropical marine yeast. Bioresour Technol 2002; 85: 35-37.

[10] Bankar AV, Kumar AR and Zinjarde SS. Environmental and industrial applications of *Yarrowia lipolytica*. Appl Microbiol Biotechnol 2009; 84: 847-865.

[11] Dominguez A, Ferminan E, Sanchez M, Gonzalez FJ, Perez-Campo FM, Garcia S, Herrero AB, San Vicente A, Cabello J, Prado M, Iglesias FJ, Choupina A, Burguillo FJ, Fernandez-Lago L, Lopez MC. Non-conventional yeasts as hosts for heterologous protein production. Int Microbiol 1998; 1: 131-142.

[12] Barth G, Gaillardin C. *Yarrowia lipolytica*. Nonconventional Yeasts in Biotechnology: A Handbook. Wolf, K. (Ed.), Springer-Berlin, Heidelberg. New York 1996; 313-388.

[13] Nicaud JM, Madzak C, Van den Broek P, Gysler C, Duboc P, Niederberger P, Gaillardin C. Protein expression and secretion in the yeast *Yarrowia lipolytica*. FEMS Yeast Res 2002; 2: 371-379.

[14] Madzak C, Treton B, Blanchin-Roland S. Strong hybrid promoters and integrative expression/secretion vectors for quasi-constitutive expression of heterologous proteins in the yeast *Yarrowia lipolytica*. J Mol Microbiol Biotechnol 2000; 2: 207-216.

[15] Margesin R, Schinner F. Bioremediation of diesel-oil-contaminated alpine soils at low temperatures. Appl Microbiol Biotechnol 1997; 47: 462-468.

[16] Zinjarde SS, Pant A. Hydrocarbon degraders from a tropical marine environment. Mar Pollut Bull 2002; 44: 118-121.

[17] Zinjarde SS, Pant A. Emulsifier from a tropical marine yeast, *Yarrowia lipolytica* NCIM 3589. J Basic Microbiol 2002; 42: 67-73.

[18]   Zógała B, Robak M, Rymowicz W, Wzientek K, Rusin M, Maruszczak J. Geoelectrical observation of *Yarrowia lipolytica* bioremediation of petrol-contaminated soil. Polish Journal Env Studies 2005; 14: 665-669.

[19]   Andreishcheva EN, Isakova EP, Sidorov NN, Abramova NB, Ushakova NA, Shaposhnikov GL, Soares MIM and Zvyagilskaya RA. Adaptation to salt stress in a salt-tolerant strain of the yeast *Yarrowia lipolytica*. Biochemistry (Moscow) 1999; 64: 1061-1067.

[20]   Zvyagilskaya R, Parchomenko O, Abramova N, Allard P, Panaretakis T, Pattison-Granberg J, Persson BL. Proton and sodium-coupled phosphate transport systems and energy status of *Yarrowia lipolytica* cells grown in acidic and alkaline conditions. J Membrane Biol 2001; 183: 39-50.

[21]   Strouhal M, Kizek R, Vacek J, Trnkova L, Nemec M. Electrochemical study of heavy metals and metallothionein in yeast *Yarrowia lipolytica*. Bioelectrochemistry 2003; 60: 29-36.

[22]   Jain MR, Zinjarde SS, Deobagkar DD, Deobagkar DN. 2, 4, 6-trinitrotoluene transformation by a tropical marine yeast, *Yarrowia lipolytica* NCIM 3589. Mar Pollut Bull 2004; 49: 783-788.

[23]   Butinar L, Santos S, Spencer-Martins I, Oren A and Gunde-Cimerman N. Yeast diversity in hypersaline habitats. FEMS Microbiol Lett 2005; 244: 229-234.

[24]   Kim SJ. Screening and its potential application of lipolytic activity from a marine environment: characterization of a novel esterase from *Yarrowia lipolytica* CL180. Appl Microbiol Biotechnol 2007; 74: 820-828.

[25]   Carreira A. Ferreira LM. Loureiro V. Production of brown tyrosine pigments by the yeast *Yarrowia lipolytica*. J Appl Microbiol 2001; 90: 372-379.

[26]   Beckerich JM, Boisramé-Baudevin A, Gaillardin C. *Yarrowia lipolytica:* a model organism for protein secretion studies. Int Microbiol 1998; 1: 123-130.

[27]   Dominguez A, Fermiñan E, Gaillardin C. *Yarrowia lipolytica:* an organism amenable to genetic manipulation as a model for analyzing dimorphism in fungi. Contrib Microbiol 2000; 5: 151-172.

[28]   Titorenko VI, Smith JJ, Szilard RK, Rachubinski RA. Peroxisome biogenesis in the yeast *Yarrowia lipolytica*. Cell Biochem Biophys 2000; 32: 21-26.

[29]   Ruiz-Herrera J and Sentandreu R. Different effectors of dimorphism in *Yarrowia lipolytica*. Arch Microbiol 2002; 178: 477-483.

[30]   Fickers P, Benetti PH, Wache Y, Marty A, Mauersberger S, Smit MS, Nicaud JM. Hydrophobic substrate utilisation by the yeast *Yarrowia lipolytica*, and its potential applications. FEMS Yeast 2005; 5: 527-543.

[31]   Beopoulos A, Chardot T, Nicaud JM. *Yarrowia lipolytica*: a model and a tool to understand the mechanisms implicated in lipid accumulation. Biochimie 2009; 91: 692-696.

[32]   Thevenieau F, Nicaud JM, Gaillardin C. Application of the non- conventional yeast *Yarrowia lipolytica*. In: Kunze SA, Satyanarayana T (eds) Diversity and potential biotechnological applications of yeasts. Elsevier Amsterdam 2008.

[33]   Szaniszlo PJ. Fungal dimorphism with emphasis on fungi pathogenic for humans. Plenum Press, New York, N. Y. 1985; 263-280.

[34]   Vanden Bossche A, Odds FC, Kerridge D. Dimorphic fungi in biology and medicine. Plenum Press, New York 1993; 105-119.

[35]   Khale-Kumar A and Deshpande MV. Possible involvement of cyclic adenosine 3',5'-monophosphate in the regulation of NADP-/NAD-glutamate dehydrogenase ratio and in yeast-mycelium transition of *Benjaminiella poitrasii*. J Bacteriol 1993; 175: 6052-6055.

[36]   McIntyre M, Breum J, Arnau J, and Nielsen J. Growth physiology and dimorphism of *Mucor circinelloides* (syn. racemosus) during submerged batch cultivation. Appl Microbiol Biotechnol 2002; 58: 495-502.

[37]   Jiménez-Bremont JF, Ruiz-Herrera J, Domínguez A. Disruption of gene YlODC reveals absolute requirement of polyamines for mycelial development in *Yarrowia lipolytica*. FEMS Yeast Res 2001; 1: 195-204.

[38]   Ogrydziak D, Bassel J, Contopoulou R and Mortimer R. Development of genetic techniques and the genetic map of the yeast *Saccharomyeopsis lipolytica*. Mol Gen Genet 1978; 163: 229-239.

[39]   Davidow LS, Apostolakos D, O'Donnelle M, Proctor AR, Ogrydziak DM, Wing RA, Stasko I, DeZeeuw JR. Integrative transformation of the yeast *Yarrowia lipolytica*. Curr Genet 1985; 10: 39-48.

[40]   Perez-Campo, FM and Dominguez A. Factors affecting the morphogenetic switch in *Yarrowia lipolytica*. Curr Microbiol 2001; 43: 429-433.

[41]   Rodríguez C, and Domínguez A. The growth characteristics of *Saccharomycopsis lipolytica:* morphology and induction of mycelial formation. Can J Microbiol 1984; 30: 605-612.

[42]   Szabo R and Stofaníková V. Presence of organic sources of nitrogen is critical for filament formation and pH-dependent morphogenesis in *Yarrowia lipolytica*. FEMS Microbiol Lett 2002; 206: 45-50.

[43]   Kim TH, Oh YS and Kim SJ. The possible involvement of the cell surface in aliphatic hydrocarbon utilization by an oil-degrading yeast *Yarrowia lipolytica* 180. J Microbiol Biotechnol 2000; 10: 333-337.

[44]   Pollack JH and Hashimoto T. The role of glucose in the pH regulation of germ-tube formation in *Candida albicans*. J Gen Microbiol 1987; 133: 415-424.

[45]   Ruiz-Herrera J, Ruiz-Medrano R, Domínguez A. Selective inhibition of cytosine-DNA methylases by polyamines. FEBS Lett 1995; 357: 192-196.

[46]   Gonzalez-Lopez CI, Ortiz-Castellanos L and Ruiz-Herrera J. The ambient pH response Rim pathway in *Yarrowia lipolytica*: identification of YlRIM9 and characterization of its role in dimorphism. Curr Microbiol 2006; 53: 8-12.

[47]   Aréchiga-Carvajal E T and Ruiz-Herrera J. The RIM101/pacC homologue from the basidiomycete *Ustilago maydis* is functional in multiple pH-sensitive Phenomena. Eukaryot Cell 2005; 4: 999-1008.

[48]   Tabor CW, Tabor H. Polyamines in microorganisms. Microbiol Rev 1985; 49: 81-99.

[49]   Takahashi T, Kakehi JI. Polyamines: ubiquitous polycations with unique roles in growth and stress responses. Ann Bot 2010; 105: 1-6.

[50]   Fuell C, Elliott KA, Hanfrey CC, Franceschetti M, Michael AJ. Polyamine biosynthetic diversity in plants and algae. Plant Physiol Biochem 2010; 48: 513-520.

[51]   Tabor CW, Tabor H. Polyamines. Annu Rev Biochem 1984; 53: 749-790.

[52]   Smitht. A. Polyamines. Annual Review of Plant Physiology 1985; 36: 117-143.

[53]   Hayashi S. In Ornithine Decarboxylase: Biology, Enzymology and Molecular Genetics (Hayashi S, ed.) PergamonPress 1989; 35-45.

[54]   Zappia V, and Pegg AE. Progress in polyamine research. Novel biochemical pharmacological and clinical aspects: advances in experimental medicine and biology, (NY: Plenum Press) Vol. 250, 1988.

[55]   Ruiz-Herrera J, Mormeneo S, Vanaclocha P, Font de Mora J, Iranzo M, Puertes I, Sentandreu R. Structural organization of the components of the cell wall from *Candida albicans*. Microbiology 1994; 140: 1513-1523.

[56]   Herrero A B, López MC, García V, Schmidt A, Spaltmann F, Ruiz- Herrera J and Dominguez A. Control of filament formation in *Candida albicans* by polyamine levels. Infect Immun 1999; 67: 4870-4878.

[57]   McCann PP, Pegg AE, Sjoerdsma A. Inhibition of Polyamine Metabolism: Biological Significance and Basis for New Therapies. Academic Press, Orlando 1987.

[58]   Shah P, Swiatlo E. A multifaceted role for polyamines in bacterial pathogens. Mol Microbiol 2008; 68: 4-16.

[59]   Igarashi K and Kashiwagi K. Polyamines: mysterious modulators of cellular functions. Biochem Biophys Res Commun 2000; 271: 559-564.

[60]   Childs AC, Mehta DJ, Gerner EW. Polyamine-dependent gene expression. Cell Mol Life Sci 2003; 60: 1394-1406.

[61]   Shabala S, Cuin TA, Pottosin I. Polyamines prevent NaCl-induced K+ efflux from pea mesophyll by blocking non-selective cation channels. FEBS Letters 2007; 581: 1993-1999.

[62]   Handa AK and Mattoo AK. Differential and functional interactions emphasize the multiple roles of polyamines in plants. Plant Physiol Biochem 2010; 48: 540-546.

[63]   Madrigal JM, Guerrero PI, Martínez IJ, Valadez B, Guzman JC, Solis E, Corona JF, Schrank A, Bremont JF, Hernandez A. Isolation, characterization and expression analysis of the ornithine decarboxylase gene (ODC1) of the entomopathogenic fungus, *Metarhizium anisopliae*. Microbiological Research 2011; 166: 494-507.

[64]   Stevens L, and Winter MD. Spermine, spermidine and putrescine in fungal development. Advances in Microbial Physiology 1979; 19, 63-148.

[65]   Inderlied CB, Cihlar RL, Sypherd PS. Regulation of ornithine decarboxylase during morphogenesis of *Mucor racemosus*. J Bacteriol 1980; 141: 696-706.

[66]   Calvo-Méndez C, Martínez-Pacheco M, Ruiz-Herrera J. Regulation of ornithine decarboxylase activity in *Mucor bacilliformis* and *Mucor rouxii*. Exp Mycol 1987; 11: 270-277.

[67]   Martínez-Pacheco M, Rodríguez G, Reyna G, Calvo-Méndez C, Ruiz-Herrera J. Inhibition of the yeast-mycelial transition and the phorogenesis of Mucorales by diamino butanone. Arch Microbiol 1989; 151: 10-14.

[68]   Martinez JP, Lopez-Ribot JL, Gil ML, Sentandreu R and Ruiz-Herrera J. Inhibition of thedimorphic transition of *Candida albicans* by the ornithine decarboxylase inhibitor 1,4 diaminobutanone: alterations in the glycoprotein composition of the cell wall. Microbiology 1990; 136: 1937-1943.

[69]   Guevara-Olvera L, Xoconostle-Cazares B, Ruiz-Herrera J. Cloning and disruption of the ornithine decarboxylase gene of *Ustilago maydis*: evidence for a role of polyamines in its dimorphic transistion. Microbiology 1997; 143: 2237–2245.

[70]   Cervantes-Chávez J A and Ruiz-Herrera J. STE11 disruption reveals the central role of a MAPK pathway in dimorphism and mating in *Yarrowia lipolytica*. FEMS Yeast Res 2006; 6: 801-815.

[71]   Cervantes-Chávez J A and Ruiz-Herrera J. The regulatory subunit of Protein Kinase A promotes hyphal growth and plays an essential role in *Yarrowia lipolytica.* FEMS Yeast Res 2007; 7: 929-940.

[72]   Cervantes-Chavez JA, Kronberg F, Passeron S and Ruiz- Herrera. Regulatory role of the PKA pathway in dimorphism and mating in *Yarrowia lipolytica.* Fungal Genet Biol 2009; 46: 390-399.

[73]   Torres-Guzman JC, Dominguez A. 1997. *HOY1,* a homeogene required for hyphal formation in *Yarrowia lipolytica.* Mol Cell Biol 1997; 17: 6283-6293.

[74]   Hurtado CA and Rachubinski RA. MHY1 encodes a C_2H_2-type zinc finger protein that promotes dimorphic transition in the yeast *Yarrowia lipolytica.* J Bacteriol 1999; 181: 3051-3057.

[75]   Hurtado CAR, Beckerich JM, Gaillardin C and Rachubinski RA. A Rac homolog is required for induction of hyphal growth in the dimorphic yeast *Yarrowia lipolytica.* J Bacteriol 2000; 182: 2376-2386.

[76]   Bassilana M, Arkowitz RA. Rac1 and Cdc42 Have Different Roles in *Candida albicans* Development. Eukaryot Cell 2006; 5: 321-329.

[77]   Barth G, Beckerich JM, Dominguez A, Kerscher S, Ogrydziak D, Titorenko V and Gaillardin. Functional genetics of *Yarrowia lipolytica.* Topics in Current Genetics 2003; 2: 227-272.

[78]   Braun BR and Johnson AD. Control of filament formation in *Candida albicans* by the transcriptional repressor TUP1. Science 1997; 277, 105-109.

[79]   Williams FE and RJ Trumbly. Characterization of TUP1, a mediator of glucose repression in *Saccharomyces cerevisiae.* Mol Cell Biol 1990; 10: 6500-6511.

[80]   Szabo R. Cla4 protein kinase is essential for filament formation and invasive growth of *Yarrowia lipolytica.* Mol Gen Genom 2001; 265: 172-179.

[81]   Leveleki L, Mahlert M, Sandrock B and Bölker M. The PAK family kinase Cla4 is required for budding and morphogenesis in *Ustilago maydis.* Mol Microbiol 2004; 54: 396-406.

[82]   Hurtado CA and RA Rachubinski. Isolation and characterization of *YlBEM1,* a gene required for cell polarization and differentiation in the dimorphic yeast *Yarrowia lipolytica.* Eukaryot Cell 2002; 1: 526-537.

[83]   Palmer GE, Johnson KJ, Ghosh S and Sturtevant J. Mutant alleles of the essential 14-3-3 gene in *Candida albicans* distinguish between growth and filamentation. Microbiology 2004; 150: 1911-1924.

CHAPTER 6

Sporothrix schenckii and General Aspects of Sporotrichosis

Leila M. Lopes-Bezerra[*] and Rosana C. Nascimento

Laboratory of Cellular Mycology and Proteomics, Biology Institute, State University of Rio de Janeiro (UERJ), Rua São Francisco Xavier 524 PHLC, 20550-013, Rio de Janeiro, Brazil

Abstract: The thermally dimorphic fungus *Sporothrix schenckii* is the etiological agent of human and animal sporotrichosis and belongs to the recently proposed *Sporothrix* complex, which includes other species based on the phylogenetic-species concept. Sporotrichosis is a deep mycosis and clinical manifestations vary from a benign lymphocutaneous form to less frequent disseminated and extracutaneous forms, mainly associated with an immunocompromised host. Animals are also susceptible to *S. schenckii* infection and cats (*Felis catus*) are known for greater predisposition to this fungal infection. Only recently has zoonotic transmission of this disease been reported in greater detail in the literature and these new data are changing the epidemiological concept of this mycosis. The gold standard for the diagnosis of sporotrichosis remains fungus isolation from biological specimens, but new diagnostic tools are under development for both human and animal sporotrichosis. The outcome of an infectious disease is not only associated with virulence factors inherent to the pathogen, but also the host immune response. Thermotolerance is described as a virulence factor for *S. schenckii* and is associated with the capacity of a clinical or environmental isolate to cause host damage. However, the description of new species and genotypes among clinical isolates and the observation that thermotolerant isolates can exhibit differences in virulence using *in vivo* models strongly supports the concept that other virulence factors could be related to the clinical manifestations or modulate the host immune response. This chapter will address new data in relation to the clinical, epidemiological and biological aspects of *S. schenckii*.

Keywords: *Sporothrix schenckii*, sporotrichosis, epidemiology, virulence factors, host-fungus interplay, clinical diagnosis, saprophytic, thermotolerance, mycosis, taxonomy, phenotypic analysis.

INTRODUCTION

In the last few decades, a significant increase has occurred in the incidence of infections caused by pathogenic fungi. This increase in the incidence of opportunistic mycoses is related to the pandemic of AIDS, the use of invasive methods in current medical practice (aggressive medical practices) or to therapeutic interventions, such as the use of immunosuppressive therapies. However, these procedures are necessary and so the incidence of fungal infections will increase accordingly. Mycoses caused by primary fungal pathogens (primary pathogens) have also been reported more frequently in the literature. Subcutaneous mycoses are more common in Latin America in regions with tropical or subtropical climates, where climatic conditions favor certain species of pathogenic fungi and, consequently, a higher number of cases are described. Many of these infections are under-reported, since they are generally benign superficial cutaneous and subcutaneous mycoses. Although sporotrichosis is the most common subcutaneous mycosis in South America, cases have been described on all continents. The significant increase in case reports of sporotrichosis in the last two decades[1] may be due to greater expertise in the clinical diagnosis of fungal infections, in general, and of sporotrichosis, in particular, or to the inclusion of fungal infections in the routine of differential diagnosis of infectious diseases that present similar clinical signs and symptoms. In the specific case of sporotrichosis, the importance of a differential

Address correspondence to Leila Lopes-Bezerra: Laboratory of Cellular Mycology and Proteomics, Biology Institute, State University of Rio de Janeiro (UERJ), Rua São Francisco Xavier 524 PHLC, 20550-013, Rio de Janeiro, Brazil. Email: leila.lopes_bezerra@pq.cnpq.br

[1]The results of searches in PubMed between 1960 and 1990 (three decades) using the key words sporotrichosis and case report obtained 260 hits, while for the last two decades (1990-2010), this number jumps to 332 hits; *i.e.,* twice the mean for case report hits per decade.

José Ruiz-Herrera (Ed)

diagnosis that includes cutaneous leishmaniasis, chromoblastomycosis, paracoccidioidomycosis, atypical mycobacteriosis and tuberculosis is indicated [1].

Although the most common clinical forms of sporotrichosis are lymphocutaneous and fixed cutaneous, there are reports in the literature of severe clinical conditions generally associated with immunocompromise status of the host or other factors that are discussed throughout this chapter. Furthermore, recent reports of clinical cases have been associated with new species of the *Sporothrix* complex, such as *Sporothix. globosa* [2, 3]. Therefore, cases of sporotrichosis caused by pathogenic species of the so-called *S. schenckii* species complex [4] should now be considered in routine clinical practice. Precise, early diagnosis of fungal infections also depends on the development of laboratory methods with high sensitivity and specificity, such as molecular methods. The accuracy of a diagnostic test is fundamental to monitoring the patient and the control of infectious diseases.

Originally, sporotrichosis was considered an occupational disease that mainly affected rural workers, florists and gardeners who were exposed to traumatic inoculation with plant debris or contaminated organic material, such that it was rarely associated with zoonotic transmission [5]. However, particularly in South America and India, cases involving peridomicile and zoonotic transmission have been reported in urban areas [6-9]. The changing epidemiological profile of the disease is associated with a higher frequency of severe clinical forms, rare or atypical for this mycosis, including fatal cases [10-13].

Briefly, new data regarding the description of new pathogenic species of the *Sporothrix* complex, as well as virulence factors inherent to the pathogen and host susceptibility may impact on the clinical and epidemiological profile of the disease.

S. schenckii and Sporotrichosis

Sporothrix schenckii was first described by Schenck in 1898, in a clinical isolate from skin lesions in a patient at The Johns Hopkins Hospital, Baltimore, USA. The organism was isolated from lesions of the index finger and forearm of a 36 year-old male who presented multiple ascending lesions and ulcerated nodules. Following isolation of the etiologic agent, the sample was sent to mycologist Erwin Smith, who concluded it was a fungus of the genus *Sporotrichum* [14]. In 1900 the disease was described for the second time, when Hektoen and Perkins isolated this pathogen from a skin lesions on the finger of a child who injured himself with a hammer, when it was classified as *Sporothrichum schenckii* [15]. In Europe, the first case was described in 1903 and more than 200 cases were reported in the following 10 years [16]. The first case of sporotrichosis in South America was described in Brazil in 1907, by Lutz and Splendore [17], when *in vitro* culture of the parasitic phase, the yeast phase, was also described.

Sporothrix schenckii belongs to Division: Ascomycota; Subdivision: Pezizomycotina; Class: Sordariomycetidae; Subclass: Sordariomycetidae; Order: Sordariales; Family: Cephalothecaceae; Genus: Sporothrix; Species: *Sporothrix schenckii*; according to the Integrated Taxonomic Information System (ITIS)[1], taxonomic serial number: 181714. Phylogenetic studies involving 127 isolates of *S. schenckii*, from different geographical origins, demonstrated that this fungal pathogen presents high genetic variability and constitutes a complex of species associated with human infections, *S. schenckii*, *Sporothrix brasiliensis* and *S. globosa* [4, 18]. Two other species of potential medical importance, *Sporothrix mexicana* and *Sporothrix albicans*, also form part of the so called *S. schenckii* species complex; however, these were not originally associated with clinical cases [4, 19].

In the last decade, several genotypic and phenotypic studies involving *S. schenckii* were conducted, revealing the existence of groups with high genotypic variability among isolates of this fungus [19-23]. In 2007, what was once known as a single species, *S. schenckii*, was proven to be several cryptic species by a combination of phenetic and genetic approaches, giving rise to the term *S. schenckii* species complex [4]. In

---

[1] Integrated Taxonomic Information System (ITIS) at http://www.itis.gov/

this complex, the species *S. schenckii*, *S. brasiliensis*, *S. globosa* and *S. albicans* show significant differences in the assimilation of sucrose, raffinose, and ribitol and growth at temperatures of 30°C, 35°C and 37°C. Two of these new species, *S. brasiliensis* and *S. globosa*, are of great interest for clinical and epidemiological reasons. In the same epidemiological area, the species *S. globosa* was recently identified as the etiologic agent in a patient with sporotrichosis lymphocutaneous [3], while *S. brasiliensis* was identified in an isolate of feline sporotrichosis [Lopes-Bezerra, unpublished results]. However, it should be highlighted that correlations between species and geographic origin, virulence and clinical form of the disease require further research.

Isolates of *S. schenckii* produce moist, membranous-looking, high colonies, with ridges or folds on the surface. Initially, colonies range in color from white to cream, becoming brown, dark gray and black. In subcultures, colonies can irreversibly lose their pigmentation, becoming white-cream. This fungus is dimorphic, such that the morphological transition is determined by temperature, although other factors may influence the dimorphic transition [24].

The saprophytic, mycelium phase is characterized by thin, hyaline, septate and branched hyphae, with slender conidiophores, and a small vesicle with sympodial arranged denticles forms on the apex, that can range from hyaline to brown. From each denticle sprouts a conidium measuring about 2 to 6 μm [24]. Conidia are gathered in floral arrangement in conidiophores, a structure characteristic of this species (Fig. 1), though they occasionally detach from these and are bilaterally arranged on the hyphae. The presence of sessile conidia, as illustrated in Fig. 1, is observed in the species *S. schenckii* (*sensu stricto*) and *S. brasiliensis*. The format of conidia varies according to species, from globular to ovoid, while some isolates present triangular conidia [4]. The parasitic, yeast phase is pleomorphic, presenting oval to cigar-shaped cells measuring about 2.5 to 5 μ m in diameter. During *in vitro* culture, the mycelium phase can be obtained at 25-28°C and the yeast phase by culture at 35-37°C [24]. When subjected to temperatures above 38.5°C, this fungus cannot grow, thus temperature is used as a form of therapy to treat skin lesions [25].

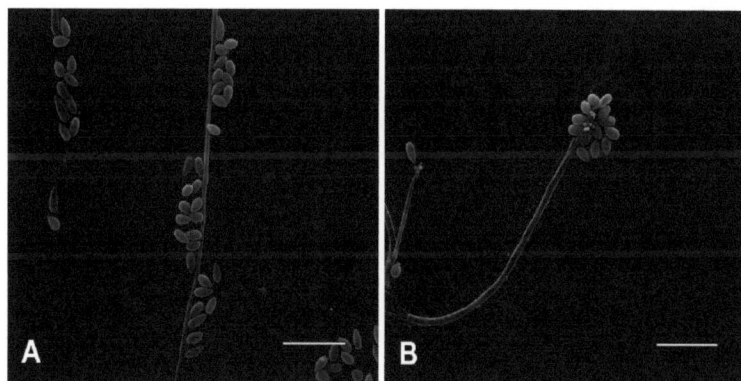

Figure 1: Scanning electron microscopy of the mycelium phase of two clinical isolates of *S. schenckii* showing sessile (A) and sympodial conidia (B); bars 10 μm (courtesy of Teixeira P.A.C. and Dr. Lopes-Bezerra L.M.).

Clinical Manifestations

In the majority of cases, sporotrichosis is a benign infection limited to the skin and subcutaneous tissue, which can disseminate to bones and internal organs. Infection occurs as a consequence of traumatic inoculation of the fungus on the skin and is often associated with lymphangitis. In certain rare cases, the disease can be primarily systemic, presenting pulmonary onset. Cutaneous disseminated and extracutaneous forms have been described in immunocompromised patients, especially those infected with the HIV virus, chronic alcoholism, diabetes and in patients who have used steroids. However, more severe forms of sporotrichosis have been reported in patients with no comorbidities. During the course of the infection, *S. schenckii* can affect the joints, presenting as osteoarthritis, while the most common clinical disorders are arthralgia and synovitis, probably due to hypersensitivity reaction [1, 26-29].

Sporotrichosis presents various clinical forms, which can be classified as: cutaneous (lymphocutaneous, fixed cutaneous and disseminated); mucosa (affecting the eyes, nose and others); extracutaneous (osteoarticular, pulmonary, meningitis, generalized); residual and special forms (erythema nodosum and erythema multiform). Precisely what determines each of these forms remains unknown and several authors have attributed this diversity to several factors, including the depth and size of the initial inoculum, thermotolerance [30], origin and genotype of the isolate [20, 21] and the host immune status [26, 27]. However, other virulence factors have been described, as detailed below. According to Rippon (1988) [14], exposure to small amounts of conidia in an endemic area may gradually confer immunity.

Although the lymphocutaneous form is the most common manifestation of the disease, involving about 80% of cases, recent data show that the more severe and atypical forms already account for about 10% of sporotrichosis cases in endemic areas mainly related to zoonotic transmission [1, 8, 13].

Lymphocutaneous

The lymphocutaneous form is the most common presentation of the disease [29]. Following the traumatic inoculation of conidia on the skin or subcutaneous tissue, a small lesion occurs and a hard papule appears between 7 and 30 days. This lesion grows slowly, with the appearance of nodules that evolve to gum, fistulize with drainage of secretions, eventually evolving to ulcers. Systemic symptoms are weak or absent. Since the disease is progressive, similar nodules will form in a chain along the lymphatic vessels, which evolve to gum and ulcerate, remaining close to the original injury or appearing in more distant regions. This chain of nodules following the lymphatic pathway, appearing after the initial skin lesion and persisting even after the open sore heals, is a classic image of sporotrichosis (ascending nodular lymphangitis). These lesions typical of the disease, known as sporotrichoid lesions, are also observed in atypical mycobacteriosis, leishmaniasis, nocardiosis, *Lupus vulgaris*, syphilis and in tuberculosis, among others [31].

Fixed Cutaneous

In this form of sporotrichosis, the lesion is restricted to the inoculation site, with no lymphatic involvement. It is characterized by hardened or verrucous plaques, with erythematous plaques and ulcers, usually located on the face, neck, torso and legs. The lesions may present spontaneous remission; however, they can recur and may persist for years if untreated [28]. According to Kauffmann *et al.* [29], strains of *S. schenckii* that cause the fixed cutaneous form of sporotrichosis tend to be more sensitive to high temperatures compared to strains that cause the lymphocutaneous form.

Cutaneous Disseminated

This form of sporotrichosis is usually associated with some type of immunosuppression in the patient or associated diseases (Fig. **2**). Certain factors, including HIV/AIDS, patients administered chemotherapy, advanced age, chronic alcoholism, diabetes, Cushing's syndrome, prolonged corticosteroid therapy, nephropathies and other disorders are predisposing conditions for the appearance of this clinical form. Following the traumatic inoculation or inhalation of conidia, dissemination occurs by hematogenous route[2] from the initial subcutaneous lesion, which may ulcerate after several weeks or even months. The lesions of this clinical form are highly varied and may appear similar to those of cutaneous tuberculosis [29].

Extracutaneous Forms

This form of sporotrichosis occurs by hematogenous propagation directly from a primary inoculation site or regional lymph nodes [28]. Hematogenous dissemination of the fungus is the most likely *via*, since evidence exists that *S. schenckii* is capable of crossing the endothelial barrier using a paracellular route [32].

This clinical manifestation of sporotrichosis is difficult to diagnose and treat and, in general, it is not one of the triage diseases considered, for example, in cases of arthritis of unknown etiology. The extracutaneous

---

[2] Cases of fungemia have rarely been reported in the literature, but hematogenous dissemination can occur.

form, like the disseminated cutaneous form, is most frequently associated with an immunocompromised clinical status [26, 27, 33].

Fatal cases reported in the literature are rare and generally due to delayed diagnosis, when performed, and low response to treatment with available antifungal agents [29, 33, 34]. The symptoms are related to the organ/tissue involved [33]. We will detail below some of the extracutaneous forms described in the literature, osteoarticular, sporotrichal meningitis and pulmonary sporotrichosis:

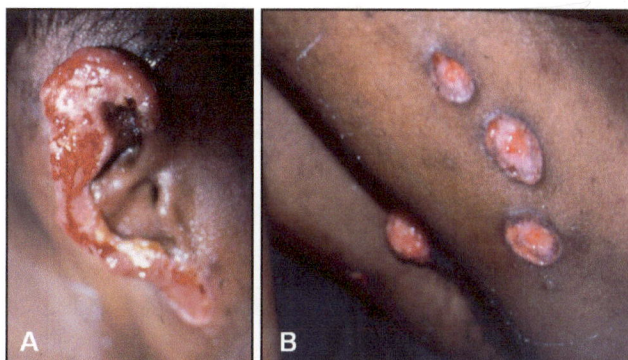

Figure 2: Disseminated sporotrichosis in a HIV patient from an endemic area of Rio de Janeiro, Brazil. The patient presented multiple lesions and osteomyelitis, including lesions on the pinna (A) and ulcers with framed edges on the thighs and between the legs, a granulomatous lesion of the scrotum (B) (courtesy of Dr. Orofino-Costa R., Mycology Laboratory of HUPE/UERJ).

Osteoarticular Sporotrichosis

Osteoarticular involvement is the most common extracutaneous manifestation of sporotrichosis [1, 28, 29, 35]. Diagnosis is usually delayed and a low immune response of patients is common [12, 36]. Osteoarticular infection can derive from the dissemination of the pathogen from skin inoculation and/or hematogenous dissemination from the lungs [29]. Osteomyelitis and cartilage destruction (Fig. **2**, panel A) are also observed in patients with the disseminated cutaneous form. The involvement of articulations can range from simple stroke or chronic synovitis to intense osteoarthritis, with total destruction of the joint. The knee joint is the area most often affected, followed by the hand, ankle and elbow [36, 37]. Arthritis may also occur as a hypersensitivity reaction [12]. This manifestation of sporotrichosis is most often associated with chronic alcoholism, patients infected with HIV or with diabetes. Clinical diagnosis is usually delayed because the symptoms mimic other more common arthritides and other pathologies, such as rheumatoid arthritis, synovitis, sarcoidosis, gout, bursitis and tuberculosis [28, 29, 34, 38].

Meningeal Sporotrichosis

Although rarely reported this is one of the most serious complications caused by this pathogen. Most cases are associated with HIV infection [29]. Cases of meningitis caused by *S. schenckii* have been described in a small number of patients and the disease progresses very slowly. Analysis of the cerebrospinal fluid (CSF) reveals increased numbers of lymphocytes, increased protein levels and decreased levels of glucose [37]. Establishing the diagnosis is difficult and the results of CSF cultures are usually negative. Treatment options are limited, with low therapeutic response and poor prognosis [29].

Pulmonary Sporotrichosis

Although a rare clinical manifestation, this form already has a number of well documented cases, sufficient to define it as a distinct clinical category. Manifestations of pulmonary infection by *S. schenckii* are similar to pulmonary tuberculosis. The typical patient is male, middle age, with chronic obstructive pulmonary disease and chronic alcoholism [29, 33]. Other conditions associated with pulmonary sporotrichosis are sarcoidosis and valvular heart disease, as well as other debilitating diseases. It is characterized by diffuse fibrosis or the formation of cavities, with an increase in mediastinal lymph nodes [39]. Diagnosis is usually delayed, leading to patient death due to late treatment of the infection or to the severity of the pulmonary disease and patients respond poorly to treatment [29].

Mucosal Sporotrichosis

Mucosal involvement is uncommon; however, it can occur and mainly affects the ocular mucosa (Fig. **3**), with granulomatous lesions accompanied by secretion and edema [1, 40], even without prior trauma. Intraocular presentation is rare and is usually a result of hematogenous dissemination of systemic lesions [40]. Even more rarely, this clinical form affects the oral cavity and larynx [41-43]. The latter can occur in the absence of cutaneous or respiratory manifestations [44]. According to Torrealba *et al.* [44], inclusion of sporotrichosis on the list of differential granulomatous fungal infections that affect the larynx is essential.

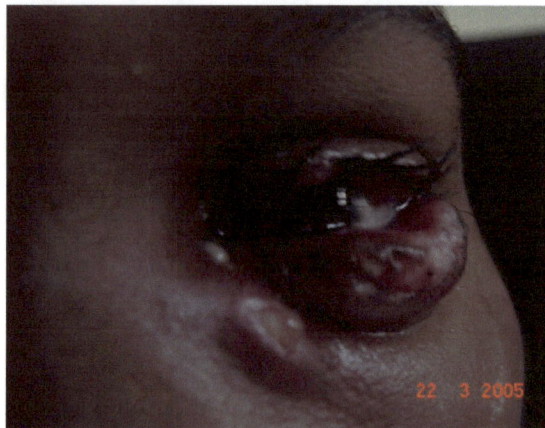

Figure 3: Mucosal sporotrichosis in an adolescent girl who reported contact with an infected cat. Ulcerated granulomatous lesion draining purulent discharge affecting the ocular conjunctiva and surrounding skin (adapted from Lopes-Bezerra *et al.* [1]).

Other Clinical Forms

In locations with a large number of cases of the disease, reports of spontaneous regression are not uncommon, nor are occurrences of hypersensitivity reactions, such as erythema nodosum or erythema multiform, even though these are considered unusual manifestations of sporotrichosis [1, 45, 46]. Generally, lesions located in the deep dermis or subcutaneous tissue result in scarring of the skin [1].

Laboratorial Diagnosis

Although new molecular methods and immunoassays for the diagnosis of sporotrichosis have been proposed in the literature, the classic mycological method is still considered the gold standard for diagnosis of this deep mycosis. Laboratorial diagnosis of sporotrichosis is achieved by isolation of *S. schenckii* from clinical samples after seeding in classic culture media, with or without the addition of antibiotics, and culture at 25 to 30°C. *S. schenckii* cultures develop from day four of incubation onward, initially as white colonies presenting a glabrous texture, which can develop pigmentation, becoming brown or black over time. Microscopically, the hyphae are hyaline and septate and the conidia are sympodial, disposed, arranged in rosettes at the apex of conidiophores or along the hyphae (Fig. **1**). Although the yeast phase presents no morphological characteristics useful for identifying the causative agent, the demonstration of dimorphism (conversion in BHI medium at 37°C) is essential for mycological diagnosis of sporotrichosis.

Sporotrichin is an antigen prepared from the culture filtrate of the mycelial or yeast phase of inactivated *S. schenckii* that is used for intradermal tests. Measurement of DTH reaction is performed after 48 hours. While a positive result is not definitive for diagnosis, when negative, it virtually excludes a diagnosis of sporotrichosis in humans, particularly the cutaneous form [47-49]. Although not used for routine diagnosis due to fake positive results, this test is often used in epidemiological studies. In veterinary medicine, despite showing strong sensitivity, intradermal reaction with sporotrichin is not used routinely because it has low specificity.

Biological specimens, such as samples of biopsy tissue, pus and exudates are used sporadically for laboratory diagnosis of sporotrichosis. Direct examination of clinical specimens, performed with KOH (20-40%) is not routinely used for diagnosis of sporotrichosis due to its low sensitivity; however, the yeast cells are occasionally analyzed by direct examination in biological specimens. When observed, the parasitic forms are best revealed using Gram staining, Giemsa, hematoxylin-eosin, PAS and Grocott-Gomori [50]. During histological examination, asteroid bodies can be observed in the tissue, which constitute globose or ovoid yeast cells enveloped by radiated eosinophilic material [51, 52]. Histopathology can be helpful in achieving the diagnosis, though the findings are generally not specific and vary according to the disease phase. Observation of inflammatory reactions and certain characteristics of the fungus in the tissue are factors that aid the laboratory technician achieve a diagnosis of sporotrichosis. Histopathological examination generally reveals that the granuloma caused by *S. schenckii* can present three distinct zones: i) a center with abscesses or necrosis (central zone); followed by ii) an area of granulomatous inflammation composed of giant epithelioid cells (tuberculoid zone); and iii) a lymphocytic halo, with granulation tissue and fibrosis (syphiloid zone) [14].

The parasitic forms are small (2-6 μm) and normally present as spherical, oval and elongated navette or cigar-shaped, without sprouting; the asteroid bodies (Splendore-Hoeppli phenomenon) can be observed [53, 54].

Some authors propose the application of immunohistochemical techniques, which show good sensitivity, in both human and animal diagnosis of sporotrichosis [50, 55]. However, the efficiency of this technique has not been proven in large-scale studies. On the other hand, the development of monoclonal antibodies against antigens of *S. schenckii* [56] could provide important advances in the application of immunohistochemical techniques.

Serological methods, such as immunodiffusion, agglutination, immunoelectrophoresis and immunofluorescence, are not routinely used to diagnose sporotrichosis in mycology laboratories, due to the lack of standardization of antigens, reagents and methodologies. Recent studies show that both a purified species-specific antigen of the cell wall of *S. schenckii* and an exoantigen preparation show high values of specificity (85-89%) and sensitivity (90-97%) in ELISA tests, with overall efficiency rates ranging from 86 to 92%, respectively [57-59]. In addition, the ELISA assays based on the detection of IgG antibodies against the antigen *Ss*CBF can be used to diagnose all clinical forms of sporotrichosis [58]. Similar results were obtained with these two types of antigenic preparations when used in ELISA assays applied to the diagnosis of feline sporotrichosis, with positive predictive values between 93% and 96% and negative predictive values between 94% and 98%, respectively [60].

Therapy

The most common protocol for the treatment of sporotrichosis is the administration of halogenated drugs (20% potassium or sodium iodide, saturated solution), in use since the 19th century. These solutions are extremely effective at treating human patients, dogs and horses.

In an era characterized by a significant increase in opportunistic fungal infections caused by fungi that until recently were considered to be saprophytic species, imidazole and triazole derivatives have emerged as promising therapeutic agents. Itraconazole is the drug recommended for the treatment of sporotrichosis [29]. Therapy with itraconazole has produced good results in the treatment of human and animal sporotrichosis, independent of the clinical form of the disease. Amphotericin B is still the drug of choice in the treatment of severe and extracutaneous forms of the disease [29, 33]. For the treatment of skin lesions in pregnant patients or those who present intolerance to the available antifungal drugs, use of local heat constitutes an adjunctive treatment for the remission of skin lesions [29].

Epidemiology and Surveillance

Sporotrichosis is a fungal infection of worldwide distribution, occurring most frequently in areas of tropical and subtropical climate, characterized by high humidity (80-95%) and moderate temperature (25-28°C). A curious variation in the incidence of this mycosis occurred in different eras, countries and continents. In the early 20th century, in Europe, the disease was considered common, particularly in France; however, it is currently considered rare on the entire continent, except for Italy and Spain [61]. In the USA, where the first cases of sporotrichosis were diagnosed, until 1932, fewer than 200 cases had been diagnosed; similarly, in South America and the Far East, there were few reports of sporotrichosis in the early 20th century [14]. Currently, higher frequencies of the disease occur in Africa, Japan, Australia, India and Central and South America, as detailed below.

Although generally acquired in isolated cases, small outbreaks of sporotrichosis can occur and when they do, they are related to a single source of contamination. One of the earliest and best known epidemics occurred in a gold mine in Witwatersrand, South Africa, between 1941 and 1944, when approximately 3,000 cases of the disease were described in miners who were infected because they handled contaminated logs used in the structure of the mine [62]. The epidemic was resolved following treatment of the wood with fungicide [63]. In the USA, the best known epidemic of sporotrichosis occurred in 1988 among forestry workers who participated in annual reforestation programs in New York and Illinois. A total of 84 cases in 15 states were identified, isolated from a single source in Wisconsin, most of them exposed to moss of the genus *Sphagnum* [64]. In a rural community in Queensland, Australia, an outbreak of sporotrichosis occurred in 1992 through human contamination by moldy hay stored in a Halloween haunted-house where parties were held [63].

As previously indicated, sporotrichosis is currently more frequently reported in other countries. In Japan, an increase in the incidence of sporotrichosis has been observed, a fact partly attributed to the improved expertise of professionals in performing the diagnosis [65]. In India, the occurrence of sporotrichosis in some regions is rare, principally the northeast. However, endemic areas do exist, like in Manipur, where 73 reported episodes of illness occurred between 1999 and 2005, and in Himachal Pradesh, where the infection is caused by contact with environmental fungal sources and the population most affected is constituted by farmers or gardeners [66-68].

For many years in Western Australia, cases of sporotrichosis were sporadic and most occurred in areas of maize cultivation; however, from the year 2000 onward, the number of cases of the disease increased. Only eight cases were reported between 1997 and 1999. In contrast, between 2000 and 2003, 41 episodes of the disease were confirmed using microbiological tests and among these, 22 (53.6%) cases occurred in the Busselton-Margaret River region, where no previous cases of the disease had been reported. In this area, the outbreak of sporotrichosis was due to the contact of rural workers with hay (82% of cases); moreover, molecular biology techniques proved that the same strain was present both in the patients' lesions and in the hay, but differed from other isolates from other regions of Australia [69-71].

Sporotrichosis is considered the most common subcutaneous mycosis in Latin America and is endemic in some regions of Mexico, Colombia, Brazil and Peru [28, 72]. The majority of reported cases occur in patients in Brazil and, in relation to other dermatoses, this mycosis showed a frequency of 0.5% in the State of São Paulo. In Albancay, Peru, hyperendemic areas occur and the incidence rate is approximately 1 case per 1,000 individuals [73]. It is still considered a rural disease by some authors, while others regard it as the most common dermatomycosis in certain urban areas, predominant in up to 93% of cases [74, 75].

Although geographical distribution of the disease is cosmopolitan, even in endemic areas, the incidence and distribution of sporotrichosis is not homogenous. Geographical distribution and the epidemiology of the disease suggest that climate, air temperature and relative humidity influence the growth of the fungus in its saprobic state. *S. schenckii* is commonly found in association with decaying plant material and is frequently isolated from the soil under certain conditions of temperature and humidity [76]. The fungus can be isolated in different climatic conditions, in cold and dry seasons, like in Mexico, or in more rainy and hot seasons, like in Uruguay, where distinct seasonal distribution of the disease occurs, with the highest frequency occurring between April and July, when the temperature varies between 13 and 20°C and the mean relative humidity is 80%. In South Africa, the fungus grows in nature in optimum temperatures between 26 and 27°C and relative humidity of 92 to 100% [77, 78]. However, in some endemic regions, the distribution of sporotrichosis is not related to seasonal differences [74, 79].

As mentioned, sporotrichosis can be acquired through traumatic inoculation of the skin by conidia or hyphae fragments following injury by contaminated organic materials, as the result of certain occupational activities, for example, fishing, hunting, gardening, horticulture, floriculture or mining. The saprophytic association of this pathogen with plants can be observed following the development of lesions caused by trauma with plant debris, including thorns, straw, twigs and slivers of wood, in approximately 10 to 60% of patients [28]. A large number of cases occur among armadillo hunters in Uruguay and although these animals are not carriers of the fungus, it can be isolated from the soil of their burrows [52, 80]. In Japan, 30% of cases of sporotrichosis involve farm workers, while in Guatemala, 45.3% of cases were reported among fishermen [65, 81]. Contamination in the laboratory may occur as a result of the manipulation of *S. schenckii* cultures [82]. Currently, new epidemiological data on zoonotic transmission in urban areas are reported and will be discussed later on.

Sporotrichosis affects individuals in all age groups; however, it is most common among young adults aged 16 to 30 years-old. Cases' occurring in individuals under 15 years of age range from 20 to 60% and the face is the region most affected in this age group. In adults, lesions more frequently occur on the arms [28, 73, 83, 84]. In regions where sporotrichosis is endemic, such as India, Japan and some Latin American countries, disease distribution between the sexes varies from region to region. In summary, the age, sex and race of the population affected by sporotrichosis play no role in the epidemiology of the disease, since the occurrence of infection is dependent on the individual's exposure to the fungus and the "entry point" of the pathogen.

Epidemiologically, the prevailing form of transmission can involve animals, as observed in Uruguay, where about 88% of cases of human sporotrichosis occur in armadillo hunters and are consequential to the scratches produced by the claws of these animals (*Dasypus* spp.), though the possibility of contagion through the earth in armadillo burrows must also be considered [52]. In Guatemala, 45.3% of cases of sporotrichosis were described among fishermen [81].

In the last few years, the incidence of sporotrichosis in urban areas has constituted a new paradigm, due to the fact that sporotrichosis in cats has gained epidemiological importance through its impact on the transmission of this mycosis to humans. The first case of natural animal transmission of sporotrichosis to humans was described in 1982 and originated from feline infection [85]. Since then successive cases have been reported in several countries, including the USA, India, Malaysia, Australia and Brazil [9, 86, 87].

Increasing incidence of feline sporotrichosis has been described in recent decades, having been considered the fourth most common deep fungal infection in the USA [88, 89]. In Brazil, sporotrichosis is the principal deep mycotic infection in cats and the incidence in dogs is increasing [90].

The spread of infection among cats appears to be related to their habits of penetrating through holes and cracks in discarded materials, rubbing themselves on the ground and up against tree bark, and due to scratches or bites when playing or fighting with other cats [85, 91]. Uncastrated male cats are more commonly affected, generally those younger than four years of age and that have unlimited access to the street. This results from the geophilic nature of the fungus and greater exposure to bites and scratches due to conflicts over territory and females [92, 93]. Greater incidence of the disease was reported in the Siamese breed [93]. In cats, clinical manifestations are more severe and the fungus tends to disseminate by hematogenous route, often leading to the animal's death [94].

The largest epidemic of sporotrichosis caused by zoonotic transmission was described in Brazil, approximately 3,244 cases in cats and 2,200 cases in humans were diagnosed between 1998 and 2009 [72]. *S. schenckii* can be isolated from the claws and oral cavities of domestic cats with sporotrichosis and even from apparently healthy animals, reinforcing the hypothesis that transmission can occur through bites or scratches from cats [90, 95]. In endemic areas in Peru, contaminated asymptomatic and infected domestic cats are currently considered a risk factor for disease transmission [7].

Although cats are now the main transmitters of zoonotic sporotrichosis, other animals are susceptible to this mycosis or play the role of reservoir. A study of cases reported in mammals, conducted between 1987 and

2007, at the Davis Veterinary Medical Teaching Hospital of the University of California, affirmed that although cats are the animals most susceptible to sporotrichosis, dogs and horses are also strongly susceptible species [96]. In contrast, the frequency of outbreaks of the disease in populations of wild and/or exotic animals is totally unknown.

Host-Fungus Interplay and Virulence Factors

Fungal infections occur when a fungal agent manages to pass through the physiological barriers that are set up to protect the host and then establishes an infection. A primary pathogen infects the host independent of any other medical condition, while an opportunistic pathogen can only cause an infection in a recipient that has already been weakened by a previous health condition. In order to establish an infectious disease, the interaction of either primary or opportunistic fungal pathogens with their host depends not only on factors inherent to the pathogen, known as virulence factors, but also upon the host condition and immune response [97]. Subcutaneous fungal infections are usually chronic conditions. They often begin after the skin has been pierced by an injury of some kind, since this allows infectious fungi to penetrate the skin and establish themselves inside the host. Eventually, a subcutaneous infection such as sporotrichosis, can evolve to severe systemic or disseminated forms, as illustrated in this chapter (Fig. **2**). In this rather complicated puzzle, the first step for a pathogen to succeed is its capacity to adhere and colonize host tissues. The binding of fungal pathogen-associated molecular patterns (PAMPs), for example, cell wall sugar polymers and proteins, to pattern recognition receptors (PRRs) on innate immune cells triggers the activation of the immune system [98, 99]. Additionally, adhesion molecules expressed on the fungus cell surface can mediate their interaction with host cells and extracellular matrix components [100].

The morphological transition and cell biology mechanisms of a dimorphic pathogenic fungus can impair the outcome of the infection or clinical manifestation of the disease. Since the cell wall is the outer cellular structure of the fungal cell, it has a key role in the host-fungus interplay. In this context, this chapter will focus on cell surface components of *S. schenckii*, which can modulate the host immune response and mediate the interaction of this pathogen with the host. Other virulence factors will also be discussed here.

Virulence Factors:

Cell Wall Components and Adhesins

The cell wall is the surface envelope of the fungal cell and plays a central role in pathogen-host interaction, thus mediating various processes associated with the pathogenesis of numerous microorganisms. Usually the fungal cell wall is mainly composed of glycoconjugates: structural polysaccharides represented by chitin and β-glucans and, cell wall glycoproteins [99]. Several types of proteins are present in the cell wall, associated or not with cell wall structural polymers like, for example, the protein families GPI-CWP (GPI-anchored cell wall proteins) and ASL-CWP (alkali-sensitive linkage-CWP proteins) [101]. Few protein or glycoprotein components have been identified so far in the cell wall of *S. schenckii*.

The first studies concerning the architecture and biochemical composition of *S. schenckii* cell wall date from the early 1970s [24]. Ultrastructural studies regarding the cell walls of both the yeast-like and mycelium phases have provided evidence that this cellular structure has several layers. The outermost layer is composed of an amorphous microfibrillar material, formerly called capsular material, which is released in the growth medium by a process known as *shedding* [24, 102, 103]. There is no strong experimental evidence that this process also occurs in other dimorphic fungi. Melanin granules are also present in the inner layers of the cell wall, as shown by transmission electron microscopy [104].

Biochemical data shows that the cell wall of *S. schencki* consists of alkali-soluble and -insoluble glucans that are identified in both morphological phases of this fungus. Alkali-soluble glucans of the yeast form of *S. schenckii* are linked by β(1,3), β(1,6) and β(1,4) bonds at 44, 28 and 28%, respectively. Insoluble glucans contain 66, 29 and 5%, respectively, of β(1,3), β(1,6) and β(1,4) bonds. No variations in β-glucan composition have been observed with the morphological transition of *S. schenckii* [105]. An immunogenic peptidorhamnonanan fraction containing 33.5% rhamnose, 57% mannose and 14.2% protein was isolated from the cell wall of the yeast phase of *S. schenckii* [106]. Peptidorhamnomannans[3] react with sera from patients with sporotrichosis and with the lectin, concanavalin A, containing the main carbohydrate epitopes recognized by human sera [57, 107-109].

In addition to rhamnose and mannose, polysaccharides containing galactose have also been identified on the surface of this fungus [110, 111]. Biochemical and immunochemical studies have shown that the O-glycosidic chains contain the main epitopes present on the cell wall of peptidorhamnomannans [1, 57, 108, 112]. Moreover, cell wall glycopeptides inhibit the adhesion of this fungus to extracellular matrix (MEC) proteins, suggesting the presence of adhesins on the surface of this pathogen [113, 114]. The ability to bind to matricial proteins is a mechanism this pathogen uses to transpose the endothelial barrier, as shown by *ex vivo* studies with human umbilical vein endothelial cell monolayers [32, 115]. Yeast cells of *S. schenckii* can bind to fibronectin, laminin and collagens type II and IV [113, 116]. A 70 kDa adhesin was recently identified and characterized either on the culture filtrate or on the cell surface of yeast cells. This protein mediates the interaction of yeast cells to fibronectin, a MEC protein present on the subendothelial matrix, and in a soluble form in plasma [56, 117]. In addition, a cell wall component of 70 kDa was shown to mediate fungus adhesion to collagenous matrices [118]. The identified gp70 antigen can also modulate the host immune response [56].

At present, few cell wall proteins and virulence factors have been characterized in this fungal pathogen of increasing epidemiological importance. This is an interesting research area, which can be improved in the near future by proteomic and genomic studies currently under development (Lopes-Bezerra L.M and Felipe M.S., unpublished results).

Proteases

The ability of fungi to invade host tissues can be also related to secreted enzymes, particularly proteases. *S. schenckii*, mainly in the yeast form of the organism, produced extracellular proteinases when cultivated in liquid media containing albumin or collagen as a nitrogen source, but not in brain heart infusion medium. Proteinase I is a 36.5 kDa serine protease inhibited by chymostatin and phenylmethylsulfonyl fluoride (PMSF), while proteinase II is a 39 kDa aspartic protease inhibited by pepstatin. The presence of both inhibitors, chymostatin and pepstatin, at 10 µg/ml in the growth medium, inhibits cell growth of *S. schenckii*; however, no effect is observed if only one inhibitor is present [118, 120]. In buffered medium at pH 6.0, the optimum pH of proteinase I, only the activity proteinase I was observed, while at pH 3.5, the optimum pH for proteinase II, only the activity of this enzyme was observed. In this case, a standard cell growth is observed independent of the pH. These observations suggest that the regulation of the production of these proteases is dependent on the environmental pH and the nitrogen source. Proteases I and II are not only expressed *in vitro*. *In vivo* studies in hairless mice have shown that these extracellular enzymes are induced in the course of infection and are recognized by antibodies in the serum of infected animals [121]. Furthermore, purified proteases I and II hydrolyze human stratum corneum, type I collagen and elastin, all natural components of the skin. However, type IV collagen, a component of basal membranes, is not a substrate. The presence of proteins with cathepsin-like activity at two days of culture and chymotrypsin-like activity at four days of culture in an exoantigen preparation of *S. schenckii* was recently described [122]. The proteolytic activity of these proteins with MW of 40 and 70 kDa, capable of hydrolyzing human IgG antibodies, was confirmed in mild acidic conditions [122].

Melanin

Melanins are macromolecules formed by oxidative polymerization of phenolic and indolic compounds. The two most important types are DHN-melanin and DOPA- melanin, each named after one of the pathway

---

[3] The formerly known peptidorhamnomannan, originally described as a single cell wall component of *S. schenckii*, is rather a complex mixture of glycopeptides and glycoproteins with a wide range of molecular weights (from 14 kDa to > 200kDa), as ascertained by SDS-PAGE and Western blot assays. The detailed chemical structure of N- and O-glycosidically linked chains of the peptidorhamnomannan is well known. For details, we recommend reference [109].

intermediates, 1,8-dihydroxynaphthalene and L-3,4-dihydroxyphenylalanine, respectively. These polymers have been implicated in both pathogenesis and cell protection in the human pathogens *Cryptococcus neoformans*, *Aspergillus fumigatus*, *Paracoccidioides brasiliensis* and *S. schenckii* [123]. Previously, it was assumed that DOPA-melanin was not synthesized by dimorphic fungi. However, to date there is strong evidence that both melanin synthetic pathways occur in *S. schenckii* [104, 124] and in other dimorphic fungi [125]. Melanin is expressed on the surface of yeast cells and conidia *in vitro*. Moreover, yeast cells can synthesize this pigment *in vivo* [126]. Melanized *S. schenckii* yeast cells are less susceptible to killing by chemically generated oxygen- and nitrogen-derived radicals than melanin-deficient cells. Melanin also interferes with yeast cell phagocytosis and diminishes the respiratory burst mediated by human monocytes and murine macrophages [127]. The availability of phenolic compounds in the culture medium increases melanin expression on the cell wall of *S. schenckii* yeast cells and, consequently, increases the virulence of this fungus [104]. Thus, the presence of melanin in the cell wall may have a protective role in this pathogen, since this pigment functions as a scavenger of free radicals.

Other

Thermotolerance has been attributed as a virulence factor of *S. schenckii* [128]. This was based on observation that clinical isolates from fixed cutaneous lesions are able to grow at 35°C, but not at 37°C. The factors responsible for thermotolerance are unknown. A recent work tried to correlate the genotype of *S. schenckii* isolates and geographical origin with thermotolerance and the clinical form of sporotrichosis [20]. Some correlation was verified among isolates from Colombia showing low thermotolerance with a higher incidence of the fixed cutaneous form of sporotrichosis. The environmental isolates from Mexico were thermotolerant [20] contrasting with previous studies showing that only 29 of 69 environmental isolates studied during an outbreak in USA grew at 37°C. These thermotolerant isolates caused fatal infections in mice [64]. The proposal of new species in the recently described *Sporothrix* complex, as denominated in this Chapter, together with their probable geographical distribution [4], could shed new light on these data in future studies.

Host Immune Response

In recent years, research related to questions of host-pathogen interaction in fungal infections have been approached with greater attention, revealing the existence of a relation between host defenses and the capacity of evasion of fungi in both human patients and animals. Moreover, the study of host response against several pathogenic fungi has gradually increased to its current level, particularly due to the increase in fungal infections diagnosed in immunocompromised patients [129].

Several clinical aspects demonstrate that cellular immunity is the main defense mechanism against fungal infections, determining the evolution of the disease. This is especially evident in two situations: i) patients presenting congenital or acquired immunodeficiency develop more severe, progressive infections; and ii) when immunocompetent patients develop severe infections, this occurs together with deficient cellular immune response, generally in association with active humoral immune response.

However, the study of infections caused by other dimorphic fungi such as *P. brasiliensis* and species of the genus *Candida*, helped clarify that components of innate and humoral immunity are involved in host defense against such pathogens, that this is related to cellular immunity and that they play a fundamental role in the immune response in fungal infection [130-132].

Sporotrichosis is usually a localized infection, affecting the skin and lymph vessels and, more rarely, the lung, bones and joints. The different clinical forms described for this mycosis could be associated with the immunological status of the host. In extracutaneous sporotrichosis, control of the infection is related to the efficiency of cellular immunity, since immunocompromised patients, those presenting a deficient cellular immune response, tend to develop the disseminated or extracutaneous forms. Not infrequently, the extracutaneous manifestation of sporotrichosis is related to infection by human immunodeficiency virus (HIV) [27, 133-135].

When the development of systemic disease occurs in animals experimentally infected with *S. schenckii*, depression of the cellular immune response is observed between weeks four and six of infection, the same period in which fungal burden in the spleen and liver of infected animals increases [136]. The transfer of spleen cells, whether previously activated or not, confers resistance to experimental infection by *S. schenckii* in immunocompetent mice [137]. Similarly, the transfer of normal thymus (from mice +/+) confers protection to congenitally athymic nude mice (nu/nu), which are more susceptible to intravenous infection with *S. schenckii* compared to heterozygous nu/+. Resistance to infection in nude mice with reconstituted thymus is related to the activation of T lymphocytes, since they present a DTH response when immunized with soluble antigens of *S. schenckii* [138].

Granulomas are formed in the host tissues in response to infection by pathogenic microorganisms and are often observed in fungal diseases. Sporotrichosis is a granulomatous mycosis and the formation of granulomas is a critical event in the immune response against *S. schenckii*, preventing tissue invasion by the pathogen. The presence of mononuclear cells in the central region surrounded by an infiltrate of activated T lymphocytes is common to the majority of infectious granulomas. Similarly, in sporotrichoid granulomas, CD4+ T lymphocytes are observed surrounding the granuloma, together with CD8+ T lymphocytes. These two populations of lymphocytes represent 1.5% of the lymphocyte population present in the sporotrichoid granuloma, in which the detection of CD83+ dendritic cells, CD68+ monocytes/macrophages and CD1a+ dendritic cells is also possible, though this last cell type is rarely verified within granulomas [139-141]. The presence of CD4+ and CD8+ T lymphocytes in sporotrichoid granulomas reinforces the hypothesis of the importance of cellular immunity in defense against this fungus; however, in experimental sporotrichosis, CD8+ T cells did not confer protection to the host, while CD4+ T cells were associated with protection against *S. schenckii* [142].

Besides lymphocytes, phagocytic cells are also involved in the control of sporotrichosis. Polymorphonuclear (PMN) leukocytes and mononuclear cells are capable of eliminating *S. schenckii* cells through the mechanisms of phagocytosis, which involve the production of oxidative metabolites, such as H_2O_2. Regarding experimental sporotrichosis, in the early stages of the granulomatous inflammatory reactions, PMN and mononuclear cells are able to contain the infection, changing the focus of the infection from microabscess to granulomatous; however, these phagocytic cells can only eliminate the fungus in the presence of activated T lymphocytes [143-145].

Macrophages play an important role in immune response as antigen-presenting cells. They are the first to recognize the infectious agent and produce a variety of effector molecules, including tumor necrosis factor-alpha (TNF-α), interleukin-1 (IL-1), IL-6 and nitric oxide (NO) [146]. The removal of microorganisms by macrophages depends on the mechanism of phagocytosis and the release of toxic agents, such as intermediate compounds of oxygen and nitrogen. The effective participation of macrophages activated by CD4+ T cells has been confirmed in experimental sporotrichosis [147]. This mechanism is important in the defense against *S. schenckii*, since it inhibits the growth of fungus in the organs of infected animals [137]. Cellular immunity deficiency in mice infected with *S. schenckii* may in part be due to low production of IL-1 and TNF-α by activated macrophages, thus impairing the amplification of the immune response [136].

The interaction of macrophages with *S. schenckii* has been shown in an animal model of chronic sporotrichosis, in which testicular nodules were formed following intraperitoneal inoculation in mice. The study verified the presence of macrophages and phagocytosed fungi, in which the nodules presented viable yeasts for six months, despite being internalized in macrophages [148]. NO production by macrophages in experimental sporotrichosis is not related to a protective immune response against the fungus, since the highest levels of NO during infection coincide with the increase in fungal load in the animals' organs and the suppression of the cellular immune response, while reduction in NO production coincides with the remission of the infection [149].

The microbicidal activity of macrophages against yeasts cells of *S. schenckii* is associated with NO production; however, no increase in fungicidal activity is observed in the presence of peroxynitrite, even though *S. schenckii* is capable of activating the oxidative burst [150]. Although NO is an important microbicidal agent in the elimination of certain pathogens, NO production during experimental infection by *S. schenckii* has a controversial role, since it contributes to a state of immunosuppression during the infection.

The set of cytokines (IL-2, IL-4, IL-10, IL-12, or TNF and interferon-gamma or IFN-γ) produced by CD4+ T cells (T helper, Th1 and Th2) determines the nature of the specific immune response to various stimuli. The polarization of the Th1 and Th2 subtypes developed from the same T cell precursor and the differentiation into two phenotypes depend on the antigen dose and costimulation for the initiation of Th differentiation [151]. The dichotomy between the role of cellular and humoral immunity against pathogens is determined in the context of the differentiation of T cell subtypes, helping to form the basis of the Th1/Th2 ratio [152]. Th1 cells produce IFN-γ, IL-2 and TNF-β, and are efficient at eliminating intracellular pathogens *via* macrophage activation. Th2 cells release IL-4, IL-5, IL-6 and IL-10 that activate humoral immunity and are secreted strongly in the presence of persistent antigens [153, 154]. In sporotrichosis, the production of IFN-γ by lymphocytes of immune mice characterizes a Th1 response capable of activating macrophages and resolving the infection in animals [142]. Furthermore, this cytokine is involved in the formation of sporotrichoid granulomas and the consequent host resistance to infection [141]. Besides IFN-γ, the Th1 response in sporotrichosis is observed by the production of IL-2 at the onset of infection. Th1 cells can serve as an initiating event through the release of IL-2 at the infection site, with subsequent activation of other Th1 cells, since this cytokine acts as a promoter of the proliferation of these cells. High production of IL-4 and IL-6 at the end of infection by *S. schenckii* can be characterized as a Th2 protection profile due to the production of high antibody titers, facilitating the remission of sporotrichotic disease [149, 155]. Given this, it appears that an effective interrelation exists between the Th1 and Th2 immune response in the resolution of sporotrichosis in an experimental model. Recently, it was observed that experimentally infected mice produce IgG1 antibodies against *S. schenckii* exoantigen. High levels of antibodies against one exoantigen, a protein of 70 kDa (gp70), can be detected in the resolution phase of the infection. Furthermore, immunization of immunocompetent and immunodeficient mice with a monoclonal antibody against gp70 was able to protect both mouse strains against infection [56, 155]. Data from the literature confirm the relevance of humoral response in the control of sporotrichosis.

CONCLUSION

Sporotrichosis is a chronic cutaneous/subcutaneous mycosis of worldwide distribution and the most frequent subcutaneous mycosis in South America. The infection is caused by the fungus *S. schenckii* and is acquired when this microorganism enters the body through a puncture wound to the skin. New species of medical importance were recently described in the *Sporothrix* complex. The disease was primary associated with occupational activities, but zoonotic transmission by infected cats has gained epidemiological importance. Knowledge concerning a variety of clinical manifestations and evidence of different genotypes associated with their geographical distribution has lead to several recent reports trying to associate these factors with specific clinical forms and fungus virulence. Therefore, a few other virulence factors were described that can impair fungus virulence. Altogether, this new information shows a complex panel that requires elucidation and the increasing epidemiological importance of sporotrichosis. We have briefly reviewed the new data in an attempt to bring new insights to this important infectious disease.

REFERENCES

[1]   Lopes-Bezerra LM, Schubach A, Costa RO. *Sporothrix schenckii* and sporotrichosis. An Acad Bras Cienc 2006;78: 293-308.

[2]   Madrid H, Cano J, Gené J, Bonifaz A, Toriello C, Guarro J. *Sporothrix globosa*, a pathogenic fungus with widespread geographical distribution. Rev Iberoam Micol 2009; 26: 218-222.

[3]   de Oliveira MM, de Almeida-Paes R, de Medeiros Muniz M, de Lima Barros MB, Galhardo MC, Zancope-Oliveira RM. Sporotrichosis caused by *Sporothrix globosa* in Rio de Janeiro, Brazil: case report. Mycopathologia 2010; 169: 359-63.

[4]   Marimon R, Cano J, Gené J, Sutton DA, Kawasaki M, Guarro J. *Sporothrix brasiliensis*, *S. globosa*, and *S. mexicana*, three new *Sporothrix* species of clinical interest. J Clin Microbiol 2007; 45: 3198-3206.

[5]   Reed KD, Moore FM, Geiger GE, Stemper ME. Zoonotic transmission of sporotrichosis: case report and review. Clin Infect Dis 1993, 16: 384-387.

[6]   Fleury RN, Taborda PR, Gupta AK, Fujita MS, Rosa PS, Weckwerth AC, *et al.* Zoonotic sporotrichosis. Transmission to humans by infected domestic cat scratching: report of four cases in São Paulo, Brazil. Int J Dermatol 2001; 40: 318-22.

[7]   Kovarik CL, Neyra E, Bustamante B. Evaluation of cats as the source of endemic sporotrichosis in Peru. Med Mycol 2008; 46: 53-56.

[8]   Schubach A, Barros MB, Wanke B. Epidemic sporotrichosis. Curr Opin Infect Dis 2008, 21: 129-33.

[9]   Yegneswaran PP, Sripathi H, Bairy I, Lonikar V, Rao R, Prabhu S. Zoonotic sporotrichosis of lymphocutaneous type in a man acquired from a domesticated feline source: report of a first case in southern Karnataka, India. Int J Dermatol 2009; 48: 1198-2000.

[10]   de Lima Barros MB, de Oliveira Schubach A, Galhardo MC, Schubach TM, dos Reis RS, Conceição MJ, do Valle AC. Sporotrichosis with widespread cutaneous lesions: report of 24 cases related to transmission by domestic cats in Rio de Janeiro, Brazil. Int J Dermatol 2003; 42: 677-681.

[11]   Barros MB, Costa DL, Schubach TM, do Valle AC, Lorenzi NP, Teixeira JL, Schubach AO. Endemic of zoonotic sporotrichosis: profile of cases in children. Pediatr Infect Dis J 2008; 27: 246-250.

[12]   Orofino-Costa R, Bóia MN, Magalhães GA, Damasco PS, Bernardes-Engemann AR, Benvenuto F, *et al.* Arthritis as a hypersensitivity reaction in a case of sporotrichosis transmitted by a sick cat: clinical and serological follow up of 13 months. Mycoses 2010; 53: 81-83.

[13]   Freitas DF, do Valle AC, de Almeida Paes R, Bastos FI, Galhardo MC. Zoonotic Sporotrichosis in Rio de Janeiro, Brazil: a protracted epidemic yet to be curbed. Clin Infect Dis 2010; 50: 453.

[14]   Rippon JW. Sporotrichosis. In: Rippon JW. Medical Mycology: The pathogenic fungi and the pathogenic actinomycetes. Philadelphia, WB Saunders, 1988; pp. 325-352.

[15]   Hektoen L, Perkins CF. Refractory subcutaneous caused by *Sporothrix schenckii*. A new pathogenic fungus. J Exp Med 1900; 5: 77-89.

[16]   Mariat F, de Bievre C. Growth characteristics and development of the content of cellular constituents of yeast and mycelial phases of *Sporotrichum schenckii*, dimorphic fungus poisonous to man. Ann Inst Pasteur (Paris) 1968; 115: 1082-1098.

[17]   Lutz A, Splendore A. Sobre uma mycose observada em homens e ratos. Rev Med Sao Paulo 1907; 21: 433-450.

[18]   Marimon R, Gené J, Cano J, Trilles L, dos Santos Lazéra M, Guarro J. Molecular phylogeny of *Sporothrix schenckii*. J Clin Microbiol 2006; 44: 3251-3256.

[19]   Arrillaga-Moncrieff I, Capilla J, Mayayo E, Marimon R, Mariné M, Gené J at al. Different virulence levels of the species of *Sporothrix* in a murine model. Clin Microbiol Infect 2009; 15: 651-655.

[20]   Mesa-Arango AC, del Rocío Reyes-Montes M, Pérez-Mejía A, Navarro-Barranco H, Souza V, Zúñiga G, Toriello C. Phenotyping and genotyping of *Sporothrix schenckii* isolates according to geographic origin and clinical form of sporotrichosis. J Clin Microbiol 2002; 40: 3004-3011.

[21]   Kong X, Xiao T, Lin J, Wang Y, Chen HD. Relationships among genotypes, virulence and clinical forms of *Sporothrix schenckii* infection. Clin Microbiol Infect 2006; 12: 1077-1081.

[22]   de Meyer EM, de Beer ZW, Summerbell RC, Moharram AM, de Hoog GS, Vismer HF, Wingfield MJ. Taxonomy and phylogeny of new wood- and soil-inhabiting *Sporothrix* species in the *Ophiostoma stenoceras-Sporothrix schenckii* complex. Mycologia 2008; 100: 647-661.

[23]   Ishizaki H, Kawasaki M, Anzawa K, Mochizuki T, Chakrabarti A, Ungpakorn R, *et al.* Mitochondrial DNA analysis of *Sporothrix schenckii* in India, Thailand, Brazil, Colombia, Guatemala and Mexico. Nippon Ishinkin Gakkai Zasshi 2009; 50: 19-26

[24]   Travassos LR, Lloyd KO. *Sporothrix schenckii* and related species of *Ceratocystis*. Microbiol Rev 1980; 44: 683-721.

[25]   Trent JT, Kirsner RS. Identifying and treating mycotic skin infections. Adv Skin Wound Care 2003; 16: 122-129.

[26]   Heller HM, Fuhrer J. Disseminated sporotrichosis in patients with AIDS: case report and review of the literature. AIDS 1991; 5: 1243-1246.

[27]   al-Tawfiq JA, Wools KK. Disseminated sporotrichosis and *Sporothrix schenckii* fungemia as the initial presentation of human immunodeficiency virus infection. Clin Infect Dis 1998; 26: 1403-1406.

[28]   Morris-Jones R. Sporotrichosis. Clin Exp Dermatol 2002; 27: 427-431.

[29]   Kauffman CA, Bustamante B, Chapman SW, Pappas PG. Infectious Diseases Society of America. Clinical practice guidelines for the management of sporotrichosis: 2007 update by the Infectious Diseases Society of America. Clin Infect Dis 2007; 45: 1255-1265.

[30]   de Albornoz MB, Mendoza M, de Torres ED. Growth temperatures of isolates of *Sporothrix schenckii* from disseminated and fixed cutaneous lesions of sporotrichosis. Mycopathologia 1986; 95: 81-83.

[31]   DiNubile MJ. Nodular lymphangitis: a distinctive clinical entity with finite etiologies. Curr Infect Dis Rep 2008; 10: 404-410.

[32]   Figueiredo CC, Deccache PM, Lopes-Bezerra LM, Morandi V. TGF-beta1 induces transendothelial migration of the pathogenic fungus *Sporothrix schenckii* by a paracellular route involving extracellular matrix proteins. Microbiology 2007; 153: 2910-21.

[33]   Kauffman CA. Sporotrichosis. Clin Infect Dis 1999; 29: 231-236.

[34]   Kohler LM, Hamdan JS, Ferrari TC. Successful treatment of a disseminated *Sporothrix schenckii* infection and *in vitro* analysis for antifungal susceptibility testing. Diagn Microbiol Infect Dis 2007; 58: 117-120.

[35]   Howell SJ, Toohey JS. Sporotrichal arthritis in south central Kansas. Clin Orthop Relat Res 1998; 346: 207-214.

[36]   Orofino-Costa R, de Mesquita KC, Damasco PS, Bernardes-Engemann AR, Dias CM, Silva IC, Lopes-Bezerra LM. Infectious arthritis as the single manifestation of sporotrichosis: serology from serum and synovial fluid samples as an aid to diagnosis. Rev Iberoam Micol 2008; 25: 54-56.

[37]   Rex JH, Okhuysen PC. Sporotrichosis. In: Mandell GL, Douglas RG, Bennett JE, eds. Principles and practice of Infectious Diseases. Local: Churchill Livingstone, 2000.

[38]   Koëter S, Jackson RW. Successful total knee arthroplasty in the presence of sporotrichal arthritis. Knee 2006; 13: 236-7.

[39]   Lupi O, Tyring SK, McGinnis MR. Tropical dermatology: fungal tropical diseases. J Am Acad Dermatol 2005; 53: 931-51, quiz 952-954.

[40]   Vieira-Dias D, Sena CM, Oréfice F, Tanure MA, Hamdan JS. Ocular and concomitant cutaneous sporotrichosis. Mycoses 1997; 40: 197-201.

[41]   Aarestrup FM, Guerra RO, Vieira BJ, Cunha RM. Oral manifestation of sporotrichosis in AIDS patients. Oral Dis 2001; 7: 134-136.

[42]   Khabie N, Boyce TG, Roberts GD, Thompson DM. Laryngeal sporotrichosis causing stridor in a young child. Int J Pediatr Otorhinolaryngol 2003; 67: 819-823.

[43]   Fontes PC, Kitakawa D, Carvalho YR, Brandão AA, Cabral LA, Almeida JD. Sporotrichosis in an HIV-positive man with oral lesions: a case report. Acta Cytol 2007; 51: 648-650.

[44]   Torrealba JR, Carvalho J, Corliss R, England D. Laryngeal granulomatous infection by *Sporothrix schenckii*. Otolaryngol Head Neck Surg 2005; 132: 339-340.

[45]   Gutierrez Galhardo MC, de Oliveira Schubach A, de Lima Barros MB, Moita Blanco TC, Cuzzi-Maya T, Pacheco Schubach TM, *et al*. Erythema nodosum associated with sporotrichosis. Int J Dermatol 2002; 41: 114-116.

[46]   Gutierrez-Galhardo MC, Barros MB, Schubach AO, Cuzzi T, Schubach TM, Lazéra MS, Valle AC. Erythema multiforme associated with sporotrichosis. J Eur Acad Dermatol Venereol 2005; 19: 507-509.

[47]   Rodrigues MT, de Resende MA. Epidemiologic skin test survey of sensitivity to paracoccidioidin, histoplasmin and sporotrichin among gold mine workers of Morro Velho Mining, Brazil. Mycopathologia 1996; 135: 89-98.

[48]   Sanchez-Aleman MA, Araiza J, Bonifaz A. Isolation and characterization of wild *Sporothrix* schenkii strains and investigation of sporototrichin reactors. Gac Med Mex 2004; 140: 507-512.

[49]   Mendez-Tovar LJ, Lemini-Lopez A, Hernandez-Hernandez F, Manzano-Gayosso P, Blancas-espinosa R, López-Martínez R. Mycoses frequency in three communities in the North mountain of the State of Puebla. Gac Med Mex 2003; 139: 118-122.

[50]   Marques ME, Coelho KI, Sotto MN, Bacchi CEJ Comparison between histochemical and immunohistochemical methods for diagnosis of sporotrichosis. Clin Pathol 1992; 45: 1089-1093.

[51]   Rodríguez G, Sarmiento L. The asteroid bodies of sporotrichosis. Am J Dermatopathol 1998; 20: 246-249.

[52]   Gezuele E, Da Rosa D. Importance of the sporotrichosis asteroid body for the rapid diagnosis of sporotrichosis. Rev Iberoam Micol 2005; 22: 147-150.

[53]   Mendez-Tovar LJ, Anides-Fonseca AE, Pena-Gonzales G, Manzano-Gayosso P, López-Martínez R, Hernández-Hernández F, Almeida-Arvizu VM. Unknown fixed cutaneous sporotrichosis. Rev Iberoam Micol 2004; 21: 150-152.

[54]   Civila ES, Bonasse J, Conti-Diaz IA, Vignale RA. Importance of the direct fresh examination in the diagnosis of cutaneous sporotrichosis. Int J Dermatol 2004; 43: 808-810.

[55]   Miranda LH, Quintella LP, Menezes RC, Dos Santos IB, Oliveira RV, Figueiredo FB, Lopes-Bezerra LM, Schubach TM. Evaluation of immunohistochemistry for the diagnosis of sporotrichosis in dogs. Vet J 2011 Jan 7. [Epub ahead of print]

[56]   Nascimento RC, Espíndola NM, Castro RA, Teixeira PA, Loureiro y Penha CV, Lopes-Bezerra LM, Almeida SR. Passive immunization with monoclonal antibody against a 70-kDa putative putative adhesin of *Sporothrix schenckii* induces protection in murine sporotrichosis. Eur J Immunol 2008; 38: 3080-3089.

[57]   Penha CV, Bezerra LM. Concanavalin A-binding cell wall antigens of *Sporothrix schenckii*: a serological study. Med Mycol 2000; 38: 1-7.

[58]   Bernardes-Engemann AR, Costa RC, Miguens BR, Penha CV, Neves E, Pereira BA, *et al.* Development of an enzyme-linked immunosorbent assay for the serodiagnosis of several clinical forms of sporotrichosis. Med Mycol 2005; 43: 487-493.

[59]   Almeida-Paes R, Pimenta MA, Monteiro PC, Nosanchuk JD, Zancopé-Oliveira RM. Immunoglobulins G, M, and A against *Sporothrix schenckii* exoantigens in patients with sporotrichosis before and during treatment with itraconazole. Clin Vaccine Immunol 2007; 14: 1149-1157.

[60]   Fernandes GF, Lopes-Bezerra LM, Bernardes-Engemann AR, Schubach TM, Dias MA, Pereira SA, de Camargo ZP. Serodiagnosis of sporotrichosis infection in cats by enzyme-linked immunosorbent assay using a specific antigen, SsCBF, and crude exoantigens. Vet Microbiol 2011; 147: 445-449.

[61]   Bachmeyer C, Buot G, Binet O, Beltzer-Garelly E, Avram A. Fixed cutaneous sporortrichosis: an unusual diagnosis in West Europe. Clin Exp Dermatol 2006; 31: 479-481.

[62]   Quintal D. Sporotrichosis infection on mines of the Witwatersrand. J Cutan Med Surg 2000; 4: 51-4.

[63]   Conias S, Wilson P. Epidemic cutaneous sporotrichosis: report of 16 cases in Queensland due to mouldy hay. Australas J Dermatol 1998; 39: 34-37.

[64]   Dixon DM, Salkin IF, Duncan RA, Hurd NJ, Haines JH, Kemma ME, Coles FB. Isolation and characterization of *Sporothrix schenckii* from clinical and environmental sources associated with the largest U.S. epidemic of sporotrichosis. J Clin Microbiol 1991; 29: 1106-1113.

[65]   Kusuhara M, Hachisuka H, Sasai Y. Statistical survey of 150 cases with sporotrichosis. Mycopathologia 1988; 102: 129-133.

[66]   Haldar N, Sharma MK, Gugnani HC. Sporotrichosis in north-east India. Mycoses 2007; 50: 201-204.

[67]   Devi KR, Devi MU, Singh TN, Devi KS, Sharma SS, Singh LR, *et al.* Emergence of sporottrichosis in Manipur. Indian J Med Microbiol 2006; 24: 216-219.

[68]   Mehta KI, Sharma NL, Kanga AK, Mahajan VK, Ranjan N. Isolation of *Sporothrix schenckii* from the environmental sources of cutaneous sporotrichosis patients in Himachal Pradesh, India: results of a pilot study. Mycoses 2007; 50: 496-501.

[69]   Beaman MH. Sporotrichosis in Western Australia. Microbiology Australia 2002; 23:16.

[70]   Feeney KT, Arthur IH, Whittle AJ, Altman SA, Speers DJ. Outbreak of sporotrichosis, Western Australia. Emerg Infect Dis 2007; 13: 1228-1231.

[71]   O'Reilly LC, Altman SA. Macrorestrition analysis of clinical and environmental isolates of *Sporothrix schenckii*. J Clin Microbiol 2006; 44: 2547-2552.

[72]   Barros MB, de Almeida Paes R, Schubach A. *Sporothrix schenckii* and sporotrichosis. Clin Microbiol Rev 2011; 24: 633-654.

[73]   Pappas PG, Tellez I, Deep AE, Nolasco D, Holgado W, Bustamante B. Sporotrichosis in Peru: description of an area of hyperendemicity. Clin Infect Dis 2000; 30: 65-70.

[74]   Macotela-Ruiz E, Nochebuena-Ramos E. Sporotrichosis among rural communities in the Northern Sierra in Puebla. Report of 55 cases September 1995 – December 2005. Gac Med Mex 2006; 142: 377-380.

[75]   Castro LGM, Salebian A. Variation in frequency of sporotrichosis at the Dermatology Clinic of "Hospital das Clínicas da Faculdade de Medicina da USP" – São Paulo, Brazil. An Bras Dermatol 1989; 64: 15-19.

[76]   Conti-Díaz IA. Epidemiology of sporotrichosis in Latin America. Mycopathologia 1989; 108: 113-116.

[77]   González-Ochoa A. Contribuciones recientes al conocimento de la esporotricosis. Gac Med Mex 1965; 95: 463-474.

[78]   Findlay GH, Vismer HF, Dreyer L. Studies on sporotrichosis. Pathogenicity and morphogenesis in the Transvaal strains of *Sporothrix schenckii*. Mycopathologia 1984; 30: 85-93.

[79]   Vismer HF, Hull PR. Prevalence, epidemiology and geographical distribution of *Sporothrix schenckii* infections in Gauteng, South Africa. Mycopathologia 1997; 137: 137-143.

[80]   Mackinnon JE, Conti-Díaz IA, Gezuele E, Civila E, da Luz S. Isolation of *Sporothrix schenckii* from nature and considerations on its pathogenicity and ecology. Sabouraudia 1969; 7: 38-45.

[81]   Mayorga R, Cáceres A, Torriello C, Gutiérrez G, Alvarez O, Ramirez ME, Mariat F. An endemic area of sporotrichosis in Guatemala. Sabouraudia 1978; 16: 185-198.

[82]   Cooper CR, Dixon DM, Salkin IF. Laboratory-acquired sporotrichosis. J Med Vet Mycol 1992; 30: 169-171.

[83]   Rubio G, Sánchez G, Porras L, Alvarado Z. Sporotrichosis: prevalence, clinical and epidemiological features Iná reference Center in Colombia. Rev Iberoam Micol 2010; 27: 75-79.

[84]   da Rosa AC, Scroferneker ML, Vettorato R, Gervini RL, Vettorato G, Weber A. Epidemiology of sporotrichosis: a study of 304 cases in Brazil. J Am Acad Dermatol 2005; 52: 451-459.

[85]   Read SI, Sperling LC. Feline sporotrichosis. Transmission to man. Arch Dermatol 1982; 118: 429-31.

[86]   Caravalho J Jr, Caldwell JB, Radford BL, Feldman AR. Feline Transmitted sporotrichosis in the southwestern United States. West J Med 1991; 154: 462-465.

[87]   Barros MB, Schubach A de O, do Valle AC, Gutierrez Galhardo MC, Conceição-Silva F, Schubach TM, *et al.* Cat-transmitted sporotrichosis epidemic in Rio de Janeiro, Brazil: description of a series of cases. Clin Infec Dis 2004; 38: 529-535.

[88]   Kier AB, Mann PC, Wagner JE. Disseminated sporotrichosis in a cat. J Am Vet Med Assoc 1979; 175: 202-204.

[89]   Dunstan RW, Reimann KA, Langham RF. Feline sporotrichosis J Am Vet Med Assoc 1986; 189: 880-883.

[90]   Schubach TM, Schubach AO, dos Reis RS, Cuzzi-Maya T, Blanco TCM, Monteiro DF *et al. Sporothrix schenckii* isolated from domestic cats with and without sporotrichosis in Rio de Janeiro, Brazil. Mycopathologia 2002; 153: 83-86.

[91]   Larsson CE, Gonçalves MA, Araujo VC, Dagli ML, Correa B, Fava Neto C. Feline sporotrichosis: clinical and zoonotic aspects. Rev Inst Med Trop Sao Paulo 1989; 31: 351-358.

[92]   Nobre MO, Castro AP, Caetano D, Souza LL, Meireles MCA, Ferreiro L. Recurrence of sporotrichosis in cats wwith zoonotic involvement. Rev Iberoam Micol 2001; 18: 1371-40.

[93]   Davies C, Troy GC. Deep Mycotic infections in cat. J Am Anim Hosp Assoc 1996; 32: 380-391.

[94]   Marques SA, Franco SR, de Camargo RM, Dias LD, Haddad V Jr, Fabris VE. Sporotrichosis of the domestic cat (*Felis catus*): human transmission. Rev Inst Med Trop Sao Paulo 1993; 35: 327-330.

[95]   Souza LL, Nascente PS, Nobre MO, Meinerz ARM, Meireles MCA. Isolation of *Sporothrix schenckii* from the nails of healthy cats. Braz J Microbiol 2006; 37: 372-374.

[96]   Crothers SL, White SD, Ihrke PJ, Affolter VK. Sporotrichosis: a retrospective evaluation of 23 cases seen in northern California (1987-2007). Vet Dermatol 2009; 20: 249-259.

[97]   Casadevall A, Pirofski LA. Virulence factors and their mechanism of action: the view from a damage-response framework. J Water Health 2009; Suppl 1: S2-S18.

[98]   Bourgeois C, Majer O, Frohner IE, Tierney L, Kuchler K. Fungal attacks on mammalian hosts: pathogen elimination requires sensing and testing. Curr Opin Microbiol 2010; 13: 401-408.

[99]   Latgé JP. Tasting the fungal cell wall. Cell Microbiol 2010; 12: 863-872.

[100] Tronchin G, Pihet M, Lopes-Bezerra LM, Bouchara JP. Adherence mechanisms in human pathogenic fungi. Med Mycol 2008; 46: 749-772.

[101] Klis FM, Brul S, De Groot PW. Covalently linked wall proteins in ascomycetous fungi. Yeast 2010; 27: 489-93.

[102] Lane JW, Garrison RG, Field MF. Ultrastrutural studies on the yeast-like and mycelia phases of *Sporotrichum schenckii*. J Bateriol 1969; 100: 1010-1019.

[103] Garrison RG, Boyd KS, Mariat F. Ultrastructural studies of the mycelium-to-yeast transformation of *Sporothrix schenckii*. J Bacteriol 1975; 124: 9599-68.

[104] Teixeira PA, de Castro RA, Ferreira FR, Cunha MM, Torres AP, Penha CV, *et al.* L-DOPA accessibility in culture medium increases melanin expression and virulence of *Sporothrix schenckii* yeast cell. Med Mycol 2010; 48: 687-95.

[105] Previato JO, Gorin PAJ, Haskins RH, Travassos LR. Soluble and insoluble glucans from different cell types of *Sporothrix schenckii*. Exp Mycol 1979; 3: 92-105.

[106] Lloyd KO, Bitoon MA. Isolation and purification of a peptide-rhamnomannan from the yeast form of *Sporothrix schenckii*. Structural and immunochemical studies. J Immunol 1971; 107: 663-671.

[107] Travassos LR, Souza W, Mendonça-Previato L, Lloyd KO. Location and biochemical nature of surface components reacting with concanavalin A in different cell types of *Sporothrix schenckii*. Exp Mycol 1977; 293-305.

[108] Lopes Alves L, Travassos LR, Previato JO, Mendonça-Previato L. Novel antigenic determinants from peptidorhamnomannans of *Sporothrix schenckii*. Glycobiology 1994; 4: 281-288.

[109] Lima OC, Bezerra LM. Identification of a concanavalin A-binding antigen of the cell surface of *Sporothrix schenckii*. J Med Vet Mycol 1997; 35: 167-172.

[110] Mendonça L, Gorin PAJ, Lloyd KO and Travassos LR. Polymorphism of *Sporothrix schenckii* surface polysaccharides as a function of morphological differentiation. Biochemistry 1976; 15: 2423-2431.

[111] Mendonça-Previato L, Gorin PA, Travassos LR. Galactose-containing polysaccharides from the human pathogens *Sporothrix schenckii* and *Ceratocystis stenoceras*. Infect Immun 1980; 29: 934-939.

[112] Lopes-Alves LM, Mendonça-Previato L, Fournet B, Degand P, Previato JO. O-glycosidically linked oligosaccharides from peptidorhamnomannans of *Sporothrix schenckii*. Glycoconj J 1992; 9: 75-81.

[113] Lima OC, Figueiredo CC, Previato JO, Mendonça-Previato L, Morandi V, Lopes Bezerra LM. Involvement of fungal cell wall components in adhesion of *Sporothrix schenckii* to human fibronectin. Infect Immun 2001; 69: 6874-6880.

[114] Lima OC, Bouchara JP, Renier G, Marot-Leblond A, Chabasse D, Lopes-Bezerra LM. Immunofluorescence and flow cytometry analysis of fibronectin nad laminin binding to *Sporothrix schenckii* yeast cells and conidia. Microb Pathog 2004; 37: 131-140.

[115] Figueiredo CC, de Lima OC, de Carvalho L, Lopes-Bezerra LM, Morandi V. The *in vitro* interaction of *Sporothrix scenckii* with human endothelial cells is modulated by cytokines and involves endothelial surface molecules. Microb Pathog 2004; 36: 177-188.

[116] Lima OC, Figueiredo CC, Pereira BA, Coelho MG, Morandi V, Lopes-Bezerra LM. Adhesion of human pathogen *Sporothrix schenckii* to several extracellular matrix proteins. Braz J Med Biol Res 1999; 32: 651-657.

[117] Teixeira PA, de Castro RA, Nascimento RC, Tronchin G, Torres AP, Lazéra M, *et al.* Cell surface expression of adhesins for fibronectin correlates with virulence in *Sporotrhix schenckii*. Microbiology 2009; 155: 3730-3738.

[118] Ruiz-Baca E, Toriello C, Perez-Torres A, Sabanero-Lopez M, Villagomez-Castro JC, Lopez-Romero E. Isolation and some properties of a glycoprotein of 70 kDa (Gp70) from the cell wall of *Sporothrix schenckii* involved in fungal adherence to dermal extracellular matrix. Med Mycol 2009; 47: 185-196.

[119] Tsuboi R, Sanada T, Takamori K, Ogawa H. Isolation and properties of extracellular proteinases from *Sporothrix schenckii*. J Bacteriol 1987; 169: 4104-4109.

[120] Tsuboi R, Sanada T, Ogawa H. Influence of culture medium pH and proteinase inhibitors on extracellular proteinase activity and cell growth of *Sporothrix schenckii*. J Clin Microbiol 1988; 26: 1431-1433.

[121] Yoshiike T, Lei PC, Komatsuzaki H, Ogawa H. Antibody raised against extracellular proteinases of *Sporothrix schenckii* in *S. schenkii* inoculated hairless mice. Mycopathologia 1993; 123: 69-73.

[122] Da Rosa D, Gezuele E, Calegari L, Goñi F. Excretion-secretion products and proteases from live *Sporothrix schenckii* yeast phase: immunological detection and cleavage of human IgG. Rev Inst Med Trop São Paulo 2009; 51: 1-7.

[123] Langfelder K, Streibel M, Jahn B, Haase G, Brakhage AA. Biosynthesis of fungal melanins and their importance for human pathogenic fungi. Fungal Genet Biol 2003; 38: 143-158

[124] Almeida-Paes R, Frases S, Fialho Monteiro PC, Gutierrez-Galhardo MC, Zancopé-Oliveira RM, Nosanchuck JD. Growth conditions influence melanization of Brazilian clinical *Sporothrix schenckii* isolates. Microbes Infect 2009; 11: 554-562.

[125] Taborda CP, da Silva MB, Nosanchuck JD, Travassos LR. Melanin as a virulence factor of *Paracoccidioides brasiliensis* and other dimorphic pathogenic fungi: a minireview. Mycopathologia 2008; 165: 331-339.

[126] Morris-Jones R, Younchim S, Gomez BL, Aisen P, Hay RJ, Nosanchuck JD, *et al.* Synthesis of melanin-like pigments by *Sporothrix schenckii in vitro* nad during mammalian infection. Infect Immun 2003; 71: 402640-33.

[127] Romero-Martinez R, Wheeler M, Guerrero-Plata A, Rico G, Torres-Guerrero H. Biosynthesis and functions of melanin in *Sporothrix schenckii*. Infect Immun 2000; 68: 3696-3703.

[128] Hogan LH, Klein BS, Levitz SM. Virulence factors of medically fungi. Clin Microbiol Rev 1996; 9: 469-488.

[129] Clemons KV, Darbonne WC, Curnutte JT, Sobel RA, Stevens DA. Experimental histoplasmosis in mice treated with anti-murine interferon-gamma antibody and in interferon-gamma gene knockout mice. Microbes Infect 2000; 9: 997-1001.

[130] Stevens DA, Domer JE, Ashman RB, Blackstock R, Brummer E. Immunomodulation in mycoses. J Med Vet Mycol 1994; 32 Suppl 1: 253265.

[131] Casadevall A, Cassone A, Bistoni F, Cutler JE, Magliani W, Murphy JW, *et al.* Antibody and/or cell-mediated immunity, protective mechanisms in fungal disease: an ongoing dilemma or an unnecessary dispute? Med Mycol 1998; 36 Suppl 1: 95-105.

[132] Calich VL, da Costa TA, Felonato M, Arruda C, Bernardino S, Loures FV, *et al.* Innate immunity to *Paracoccidioides brasiliensis* infection. Mycopathologia 2008, 165: 223-236.

[133] Plouffe JF Jr, Silva J Jr, Fekety R, Reinhalter E, Brownw R. Cell mediated immune responses in sporotrichosis. J Infect Dis 1979; 139: 152-157.

[134] Gori S, Lupetti A, Moscato G, Parenti M, Lofaro A. Pulmonary sporotrichosis with hyphae in a human immunodeficiency virus-infected patient. A case report. Acta Cytol 1997; 41: 519-521.

[135] de Araujo T, Marques AC, Kerdel F. Sporotrichosis. Int J Dermatol 2001; 40: 737-742.

[136] Carlos IZ, Zini MM, Sgarbi DB, Angluster J, Alviano CS, Silva CL. Disturbances in the production of interleukin-1 nad tumor necrosis factor in disseminated murine sporotrichosis. Mycopathologia 1994; 127: 189-194.

[137] Shiraishi A, Nakagaki K, Arai T. Role of cell-mediated immunity in the resistance to experimental sporotrichosis in mice. Mycopathologia 1992; 120: 15-21.

[138] Dickerson CL, Taylor RL, Drutz DJ. Susceptibility of congenitally athymic (nude) mice to sporotrichosis. Infect Immun 1983; 40: 417-420.

[139] Tanuma H, Asai T, Abe M, Nishiyama S, Katsuoka K. Case report. Lymphatic vessel-type sporotrichosis: immunohistochemical evaluation and cytokine expression pattern. Mycoses 2001; 44: 316-320.

[140] Koga T, Duan H, Urabe K, Furue M. Immunohistochemical localization of activated and mature CD83+ dendritic cells in granulomas of sporotrichosis. Eur J Dermatol 2001; 11: 527-529.

[141] Koga T, Duan H, Furue M. Immunohistochemical detection of interferon-gamma producing cells in granuloma formation of sporotrichosis. Med Mycol 2002; 40: 111-114.

[142] Tachibana T, Matsuyama T, Mitsuyama M. Involvement of CD4+ T cells and macrophages in acquired protection against infection with *Sporothrix schenckii* in mice. Med Mycol 1999; 37: 397-404.

[143] Cunningham KM, Bulmer GS, Rhoades ER. Phagocytosis and intracellular fate of *Sporothrix schenckii*. J Infect Dis 1979; 140: 815-817.

[144] Rex JH, Bennet JE. Administration of potassium iodide to normal volunteers does not increase killing of *Sporothrix schenckii* by their neutrophils or monocytes. J Med Vet Mycol 1990; 28: 185-189.

[145] Miyaji M, Nishimura K. Defensive role of granuloma against *Sporothrix schenckii* infection. Mycopathologia 1982; 80: 117-124.

[146] Sorimachi K, Akimoto K, Hattori Y, Ieiri T, Niwa A. Secretion of TNF-alpha, IL-8 and nitric oxide by macrophages activated with polyanions, and involvement of interferon-gamma in the regulation of cytokine secretion. Cytokine 1999; 11: 571-578.

[147] Silva AC, Bezerra LM, Aguiar TS, Tavares D, Araujo LM, Pinto CE, Ribeiro OG. Effect of genetic modifications by selection for immunological tolerance on fungus infection in mice. Microbes Infect 2001; 3: 215-222.

[148] Hiruma N, Yamaji K, Shimizu T, Ohata H, Kukita A. Ultrastructural stdy of tissue reaction of mice against *Sporothrix schenckii* infection. Arch Dermatol Res 1988; 280: S94-S100

[149] Maia DC, Sassá MF, Placeres MC, Carlos IZ. Influence of Th1/Th2 cytokines and nitric oxide in murine systemic infection induced by *Sporothrix schenckii*. Mycopathologia 2006; 161: 11-19.

[150] Fernandes KS, Coelho AL, Lopes-Bezera LM, Barja-Fidalgo C. Virulence of *Sporothrix schenckii* conidia and yeast cells, and their susceptibility to nitric oxide. Immunology 2000; 101: 563-9.

[151] Constant SL, Bottomly K. Induction of Th1 and Th2 CD4+ T cell response: the alternative approache. Annu Rev Immunol 1997; 15: 297-322.

[152] Mosmann TR, Coffman RL. TH1 and TH2 cells: different pattern of lymphokine secretion lead to different functional properties. Annu Rev Immunol 1989; 7: 145-173.

[153] Mosmann TR, Cherwinski H, Bond MW, Giedlin MA, Coffman RL. Two types of murine helper T cell clone. I. Definition according to profiles of lymphokine activities and secreted proteins. J Immunol 1986; 36: 2348-2357.

[154] Rengarajan J, Szabo SJ, Glimcher LH. Transcriptional regulation of Th1/Th2 polarization. Immunol Today 2000; 21: 479-483.

[155] Nascimento RC, Almeida SR. Humoral immune response against soluble and fractionate antigens in experimental sporotrichosis. FEMS Immunol Med Microbiol 2005; 43: 241-247.

CHAPTER 7

Yeast-Hypha Dimorphism in Zygomycetous Fungi

V. Ghormade[1], E. Pathan[2] and M. V. Deshpande[2],*

[1]Centre for Nanobioscience, Agharkar Research Institute, Pune 411004, India and [2]Biochemical Sciences Division, National Chemical Laboratory, Pune 411008, India

Abstract: The zygomycetous fungi occupy many important niches with impact on humans. To list a few, biocatalysts to produce steroids, organic acids, β-carotene, a variety of fermentation products, sources for chitosan and other polymers, biocontrol of insect pest and causative agents of mycoses and storage decays. The zygomycetes are also one of the important tools for recombinant studies. This group consists of ten orders, 32 families, around 124 genera, and 870 species. Earlier, a distinct characteristic *viz.* the reversible transition between the yeast and hyphal form, was reported in genera *Mucor*, *Cokeromyces* and *Mycotypha*. Similar transitions or giant cell formation now have been reported in other zygomycetes such as *Benjaminiella*, *Rhizopus*, *Absidia*, *Chytridiopsis* and others. The environmental perturbations triggering vegetative differentiation in different genera and cause-effect relationship of enzymes involved in ammonia assimilation, cell wall synthesis and degradation and other pathways will be discussed. Furthermore, to understand the molecular mechanism underlying this phenomenon the role of different genes such as ornithine decarboxylase, glutamate dehydrogenase, chitin synthase, actin, tubulin and others will be explored. The issues regarding the positioning of dimorphic zygomycetes in the evolution of fungi will also be addressed. The possibility of the use of dimorphism of zygomycetes for various applications will also be given.

Keywords: *Benjaminiella*, chitin deacetylase, chitin synthase, glutamate dehydrogenase, *Mucor*, ornithine decarboxylase, polyamines, yeast-hypha transition, zygomycota, dimorphism, differentiation cell wall.

INTRODUCTION

Zygomycetes and Trichomycetes are the two classes of phylum Zygomycota. The members of these classes are considered primitive as they lack well defined fruiting structures and have coenocytic hyphae [1]. Other morphological differentiating structures are asexual sporangiospores, arthrospores, chlamydospores, and in some species unicellular yeast cells. Zygomycetous fungi are characterized by a sexual zygospore. Kirk *et al.* [2] in the Dictionary of the Fungi described the Zygomycetes as consisting of 10 orders, 32 families, 124 genera, and 870 species. Among the zygomycetes, *Mucor* and *Rhizopus* significantly contribute in the industrial activities from citric acid production to a variety of biotransformation reactions using lipases. This is a second largest class which contributes in steroid transformations. One of the important fungal cell wall polymer, chitosan can be commercially produced using fungal biomass. *Absidia coerulea*, *Rhizopus delemar*, *Benjaminiella poitrasii*, *Cunnighamella blackesleeana* and *Mortierella isabelina* have high contents of chitosan and therefore have commercial significance for the chitosan production.

Though it is a rare phenomenon, zygomycetous fungi also exhibit a dimorphic, unicellular yeast or filamentous hyphal growth making them useful model system to study eukaryotic differentiation processes. Earlier, this distinct characteristic *viz.* the reversible transition between the yeast and hyphal form, was reported in genera *Mucor*, *Cokeromyces* and *Mycotypha*. Similar transitions or giant cell formation now have been reported in genera such as *Benjaminiella*, *Rhizopus*, *Absidia*, *Chytridiopsis* and others. The subsequent sections will describe the biochemical and molecular mechanisms studied in different dimorphic zygomycetous fungi and their possible applications.

Mucor

Mucor was first observed by Robert Hooke in 1665 then the morphological variability in *Mucor* was noted

*Address correspondence to Mukund V. Desphande: Biochemical Sciences Division, National Chemical Laboratory, Pune 411008, India. Email: mv.deshpande@ncl.res.in

José Ruiz-Herrera (Ed)

in 1838 and Louis Pasteur reported fermentative capability of *Mucor* and described dimorphic habit of *M. racemosus* in 1876 [3]. This is, however, one characteristic which makes several, but not all, *Mucor* sp. quite distinct from other zygomycetes. Examples of dimorphic *Mucor* sp. would include *M. racemosus*, *M. rouxii*, *M. genevensis*, *M. bacilliformis*, *M. mucedo*, *M. hiemalis*, *M. circinelloides* and *M. subtilissimus* (certain strains) [3-6]. While in case of *M. miehei*, *M. pusillus*, and *M. rammanianus*, this phenomenon is not established yet [7].

The environmental perturbations with respect to the temperature, pH, and nutrient are usual triggers for dimorphism, Yeast-Hypha, change in a number of fungi. In case of *Mucor* species anaerobiosis, *viz.* absence or low oxygen and/or presence of either CO_2 or N_2 triggered yeast development [8]. The presence of hexose was also critical for the fermentation and the yeast development. The inhibitors of oxidative phosphorylation or respiration such as potassium cyanide, antimycin A and oligomycin help organism to grow in a yeast morphology [9]. In the presence of phenethyl alcohol (PEA), inhibitor of RNA and protein synthesis and capable of altering mitochondrial properties, sporangiospores of *M. rouxii* germinated in the yeast-like form cells [10]. *M. subtilissimus* and *M. genevensis* also responded similarly to the presence of PEA. Similar to PEA, in case of *M. fragilis* ethanol addition triggered spore germination in the yeast-form by changing the unsaturated/saturated fatty acid ratio [11]. The additives like myoinositol or Zn^{++} in the medium induced yeast development in *M. circinelloides* suggesting their role as second messengers in signal transduction and subsequently in yeast induction [5].

Mycotypha

Two genera, *Mycotypha* and *Benjaminiella* are classified in the family mycotyphaceae [12]. Under *Mycotypha* three species, namely *M. africana*, *M.indica* and *M. microspora* have been described. Most of the members were isolated from dung, soil and plant material. Number of studies were carried out regarding dimorphism in *Mycotypha* [13-15]. The yeast-form was triggered in *M. africana* and in *M. microspora* by anaerobiosis, increase in temperature, or pH and with high glucose in the medium [14-15]. An *M. indica* strain isolated from turkey dung exhibited yeast-form development on nutrient- rich media [16], addition of PEA, cycoheximide, chloramphenicol, antimycin A or oligomycin in the growth medium also triggered yeast-form in *Mycotypha* [15], an effect that agrees with the isolation of a respiratory-deficient mutant which remained in a monomorphic yeast-form.

Cokeromyces

C. recurvatus was first isolated in 1950 from lizard and rodent dung and also from soil samples. *C. recurvatus* was also reported in clinical samples including endocervical specimens, vaginal secretions, urine, pleural/peritoneal fluid and abscess fluid [17-19]. In most of the tissues the organism was seen in yeast form. When *C. recurvatus* was grown on brain heart infusion agar with sheep blood at 37 °C it produced yeast cells while on potato dextrose agar the asexual sporangiospores and zygospores were seen at 25° C [19]. The organism produced yeast-form cells at 37° C under anaerobic condition [18].

Benjaminiella

The genus *Benjaminiella* has suffered many taxonomical transfers, specially a type species *B. poitrasii*, was originally described as a species of *Cokeromyces* and later it was transferred to *Mycotypha*. Therefore, number of observations made earlier for *C. poitrasii* or *M. poitrasii* have been summarized under this genus. Price *et al.* [13] yeast – growth in *C. poitrasii* in presence of PEA and high glucose in the medium. *Cole et al.* [20] observed that high glucose and organic nitrogen presence enhanced H-Y transition in *M. poitrasii*. Other than *B. poitrasii* two other species, *B. multispora* and *B. youngii* have been reported. All the three species form yeast-like cells in the nutrient- rich media. Kirk [21] reported that sporangiospores from *B. youngii* isolated from the lizard dung, on malt agar initially formed colonies of yeast-like cells. Same was reported in case of *B. poitrasii* indicating role of organic nitrogen in supporting yeast form growth [22]. *B. poitrasii* has been extensively studied to understand biochemical correlates for yeast-hypha transition [22-24]. The main dimorphism triggering conditions reported were changes in temperature, pH, glucose concentration and the presence of zinc ions. Unlike *M. fragilis*, ethanol induced the hyphal development in *B. poitrasii* [11, 22].

Rhizopus, Absidia, Chytridiopsis and Others

Rhizopus oryzae was reported to exhibit temperature dependent H-Y transition. Karmakar and Ray [25] reported higher endoglucanase production at 37°C in the yeast form cells than in the hyphal cells.

Organisms of the class *Zygomycetes* were first reported in 1800 to cause disease in humans. This first report was for *Absidia corymbifera* in a cancer patient [26]. *A. corymbifera* and *Absidia blakesleeana* were reported to form giant cells which were pleomorphic, oval to irregular shape, intercalated or in clusters in potato dextrose cultures stored under mineral oil for 3-44 years [27]. However, it was not clear whether the giant cell formation could be attributed to the anaerobic condition developed due to mineral oil layer.

In the case of *Basidiobolus ranarum* the fluorescence brightener Rylux BSU (RBSU) inhibited the polarized growth of hyphae and induced isotropic growth resulting in spherical thick-walled cells [28]. This induced H-Y transition was due to the affinity of the fluorescent dye towards the cell wall polymers and its accumulation inhibited hyphal tip growth. The chitosan contents were reported to be increased and thus the artificial H-Y transition. Similar effect was also reported in *Saccharomyces cerevisiae* [29]. Mycelial segmentation released rounded cells of *B. ranarum* and of *B. haptosporus* in synthetic media containing glucose and ammonium salts [30]. The rounded cells (formed by a process called "darmform morphoenesis") were found to secrete higher level of alkaline protease than the hyphal cells.

Conidiobolus an insect pathogenic fungus produced mycelium and also variably shaped hyphal bodies by pinching off or dissociation of intercalary cells devoid of cytoplasm [31], while Tonka *et al.* [32] reported a distinct phenomenon of multiple thin-walled developmental stages of *Chytridiopsis typographi* in the epithelial cells of the host, spruce bark beetle.

Morphological Mutants

It was indeed possible to isolate morphological mutants which under dimorphism triggering conditions for the respective organism exhibited no change in the morphology [22, 33]. To understand possible biochemical correlates of dimorphism this was one of the approaches used. Storck and Morrill [34] isolated respiration deficient mutants of *M. bacilliformis* which remained in yeast form when grown in the presence of oxygen. Later Ruiz-Herrera *et al.* [33] reported isolation and biochemical characterization of fourteen yeast form mutants of *M. bacilliformis*. It was further reported that all mutants were defective at some point in oxidative metabolism and had low levels of ornithine decarboxylase, but did not show appreciable correlation with cyclic AMP and glutamate dehydrogenase levels. On the other hand Khale *et al.* [22] reported isolation of two yeast form and one hyphal form monomorphic mutants of *B. poitrasii* which explicitly showed correlation with cAMP level and glutamate dehydrogenase activities, and also confirmed the role of chitin metabolism in the dimorphism [22, 35, 36].

Earlier Bartnicki-Garcia and Nickerson [37] reported the quantitative differences in mannan contents of yeast and hyphal form of *M. rouxii* indicating its possible role in the morphological change. Similar quantitative differences in the manose contents of yeast and hyphal form of *B. poitrasii* were also reported [38]. However, the yeast form monomorphic mutants showed mannose contents closer to the hyphal form than the parent yeast form cells. While chitin/chitosan contents of both parent and mutant yeast forms and mutant were similar indicating possible cause-effect relationship between chitin metabolism and morphogenesis in *B. poitrasii*.

Cell Wall: a Morphology Determinant

Although the environmental conditions triggering yeast to hypha transition vary in different fungi the molecular and biochemical aspects of dimorphism are apparently similar. Till now different biochemical correlates, enzymes of carbon and nitrogen pathways, cell wall metabolism, polyamine synthesis, sulfur metabolism, intracellular effectors like cAMP and Ca-calmodulin have been extensively studied [3, 35, 39].

The biochemical and molecular changes which show correlation with Y-H transition eventually narrow down to the cell wall metabolism and in turn cell wall polymer deposition pattern [3, 39-41]. Indeed all the fungal species exhibit three cell wall deposition patterns in their life cycle [42]. This is non-polarized in the yeast form, and polarized during hyphal tip growth and occasionally deregulated polarized deposition pattern for irregular shapes.

Isolation of protoplasts from yeast and hyphal form cells and their regeneration was reported to be a useful approach to understand the relative importance of the cell wall polymers to decide the morphological outcome [39-40, 43]. As chitosan is the major structural component of the cell walls of zygomycetous fungi, the protoplasts from *M. rouxii* [44, 45], *M. racemosus* [46], *Mortierella ramanniana* [47], *Rhizopus nigricans* [48], *Rhizopus* and *Conidiobolus* species [40], *Absidia glauca* [49], and *Phycomyces* sp. [50] were obtained using an enzyme mixture containing mainly chitosanase. In the fungi which exhibit dimorphism, formation of an irregular mass mainly of chitin in the initial phase of regeneration may be a pre-requisite to determine the morphological either yeast or hyphal outcome [39].

The fungal cell wall structure, growth and morphogenesis have been extensively reviewed by different researchers [51-56]. Alterations, both qualitative and quantitative in the chemical composition of cell wall are important in the morphological changes occurring during the Y-H transition. The major cell wall polysaccharides of the zygomycetous fungi are made of glucosamine [57], and it was reported that glucosaminoglycans existed in complex with glucans and were responsible for hyphal morphogenesis [51, 58]. Domek and Borgia [59] demonstrated that yeast cells of *M. racemosus* showed a relatively low differential rate of chitin-plus chitosan synthesis and that was three fold higher in hyphal cells. In presence of dibutyryl cyclic adenosine monophosphate (dbcAMP) that favoured the yeast form, aerobically grown cells displayed a lowered rate of chitin-plus-chitosan synthesis. Furthermore, it was suggested that there was considerable modification of the cell wall polymers in yeast cells *in vivo* that led to increased chitosan content. Presence of chitosan may confer extra plasticity needed for the spherical yeast cell walls and the glucosamine moiety in the polymer may bind to the matrix components of the cell walls [59]. In *B. poitrasii* the chitin contents of the hyphal form were 3 fold higher than the yeast form [38]. It was further reported that glucosaminoglycans and their cross links with the glucans could be an important contributors in the morphological transitions in *B. poitrasii*.

Among the carbohydrates found in fungal cell walls, mannans were suggested to be correlated with the morphological change [60]. The anaerobically grown yeast-form cell walls of *M. rouxii* contained 5 fold more mannan than the aerobically grown hyphal cell walls [60]. Based on amino acid analysis of cell walls in *Mucor* (where the yeast wall is rich in aspartic acid), Bartnicki-Garcia [61] suggested that incorporation of an aspartic acid rich mannoprotein into the cell wall could be responsible for the alteration in its formation, resulting in the morphological change. However, in *M. poitrasii,* the concentration of this amino acid was found to be higher in the hyphal form than in the yeast form cells [20]. A similar observation was reported for *B. poitrasii* [38].

Biochemical And Molecular Correlates of Morphogenesis

Carbon Metabolism

The carbon metabolism with respect to morphogenesis has been extensively studied in case of *Mucor* species [3, 57, 62-64]. The studies, however, were limited due to the different responses of dimorphic species of *Mucor* to the critical environmental factors. For instance, anaerobiosis with high CO_2 atmosphere and high concentration of hexose were necessary for the yeast development in *M. rouxii*. On the other hand, *M. genevensis* showed yeast growth at high hexose concentrations even under aerobic condition. *M. racemosus*, in contrast, did not require presence of CO_2 or a high glucose concentration to produce yeast form under anaerobic conditions.

The yeast growth of *M. racemosus* required the presence of a hexose in the growth medium, whereas hyphal growth occurred with a variety of carbon sources.

This implies that a relationship exists between hexose catabolism, fermentation rates and yeast morphology. It was reported in case of *Mucor* that 14% of glucose was catabolised *via* pentose phosphate pathway (PPP) and remaining by Embden-Meyerhof-Parnas (EMP) pathway in yeast cells, while 28% was processed *via* PPP in hyphal cells. This shift in the route could be correlated with the morphology [64]. Three major control points in the catabolism of hexose, namely hexokinase (EC 2.7.1.1), phosphofructokinase (EC 2.7.1.56), and pyruvate kinase (EC 2.7.1.40) were suggested to opperate in the fungus [63]. It was reported that two isozymes of pyruvate kinase isolated from *M. racemosus* were not directly related to the morphology *per se,* but were dependent on the presence or absence of glucose [63]. Inderlied and Sypherd [64] reported two additional enzymes, glucose-6-phosphate dehydrogenase (EC 1.1.1.49) and 6-phosphogluconate dehydrogenase (EC 1.1.1.44) which possibly could have correlation with the morphology. Barrera and Corral [62] observed increased activity of pyruvate decarboxylase (EC 4.1.1.1), an enzyme involved in pyruvate to acetaldehyde and CO_2 conversion, in yeast-like cells compared to the hyphal form cells of *M. rouxii.* However, the correlation of activity levels and morphology was not evident.

The transition of *C.poitrasii* from fermentative to oxidative metabolism with limited glucose was reported [65] by Rogers and Gleason [64]. When oxygen concentration increased the fermentative to oxidative metabolism, the morphological change was seen. However, unlike *M.rouxii* and *M. genevensis* there was no morphological shift *viz.* yeast- hyphal form.

In case of *B.poitrasii*, it was reported that high concentration of glucose and complex nitrogen favoured the growth of the yeast form, while the organism exhibited hyphal morphology on a variety of carbon sources such as xylose, mannose, sucrose, lactose, starch and carboxy methyl cellulose [21]. The alcohol dehydrogease activity was detected in the yeast-form cells. However, in case of *M. circinelloides* NAD-dependent alcohol dehydrogenase (ADH) activity was detected in aerobically grown hyphal cells and in anaerobically grown yeast cells although it became higher in these, at a later period [66].

Nitrogen Metabolism

It is well documented fact that in fungi the assimilation of ammonium into glutamate and glutamine plays a key role in the nitrogen metabolism [67-70]. The enzymes involved are: NAD- and NADP-dependent glutamate dehydrogenases (NAD-GDH, EC 1.4.1.2 and NADP-GDH, EC 1.4.1.4) and/or glutamate synthase (GOGAT, EC 1.4.1.14) and glutamine synthetase (GS, EC 6.3.1.2). In fungi such as, *S. cerevisiae* [71]. *M. racemosus* [72], *Candida boidinii* [73] and *Candida utilis* [74] the NADP-GDH was the major route both in conditions of excess and limited nitrogen than GS/GOGAT. While in *Candida tropicalis, Candida parapsilosis* and *Candida albicans* the GS/GOGAT pathway was reported to be important for ammonia assimilation [75].

The *in vivo* regulation of GDHs was studied in *M. racemosus* during the Y-H transition [3, 72]. The effect of carbon and nitrogen source on the activity of NAD-GDH suggested that its role was catabolic while NADP-GDH activities were anabolic. The NAD-GDH activity was greater in hyphal cells than in yeast form cells grown on the same medium. During Y-H transition the increase in NAD-dependent activity preceded the appearance of hyphal cells both under aerobic and anaerobic conditions. Exogenously added dbcAMP prevented the increase in NAD-dependent GDH concomitantly with the suppression of morphological differentiation. The NADP-dependent activity did not change appreciably during transition. Ruiz-Herrera *et al.* [33] using yeast monomorphic mutants of *M. bacilliformis* showed that unlike *M. racemosus* there was no significant correlation between NAD-GDH activity and morphology.

In *B. poitrasii*, NAD-GDH enzyme levels were found to be almost 10 fold lower in the yeast form cells than the hyphal form, whereas NADP-GDH activities were found to be 7 fold higher in the yeast form cells [35]. Glucose was reported to repress NAD-GDH and induce NADP-GDH. Since both NAD- and NADP-dependent GDH activities were found in *B. poitrasii*, the quantitative relationship between these two enzymes expressed as the NADP-GDH/NAD-GDH activity ratio (GDH ratio) and the morphology was suggested to be an important morphogenetic factor [35]. For instance H-Y transition was preceded by the increase in the GDH ratio while reverse was true during Y-H transition. The monomorphic yeast-form mutants showed a high GDH ratio and maintained the yeast morphology. Other two enzymes, GS and GOGAT, did not show any appreciable correlation with morphology of *B. poitrasii*. It was also reported that there were two NADP-GDH enzymes, one expressed in yeast and another in hypha and one NAD-GDH expressed in both forms [67]. The NADP-GDH isozymes showed distinct characteristics of temperature stability, optimum temperature, Km towards α-ketoglutarate as a substrate.

Fructose 6-phosphate (F6P), a common precursor in the synthesis of two cell wall polymers, mannan and chitin, was regulated by NAD- and NADP-GHDs in *S. cerevisiae* [76-78]. Osmond *et al.* [79] reported that when chitin synthesis pathway was blocked, mannan synthesis increased and *vice versa*. GDHs reversibly converted glutamate and α-ketoglutarate [78]. In *B. poitrasii*, the chitin contents in yeast, hyphae, and a monomorphic yeast-like mutant were correlated with the relative proportion of NAD- and NADP-GDH activities [35]. Joshi *et al.* [80] suggested that the presence of α-KGA or isophthalic acid repressed or inhibited NAD-GDH in *B. poitrasii*, leading to decreased input of glutamine to the chitin pathway which resulted in a decreased cell wall chitin content. Furthermore, the decrease in cell wall chitin was compensated by the increased flux of F6P towards mannan synthesis leading to enhanced mannan content in the cell walls which added to the roughness of the cell surface. Increased cell surface roughness correlated with the faster flocculation (due to increased mannan-lectin interactions). Conversely, addition of glutamate resulted in increased NAD-GDH activity, increased chitin content, decreased mannan content and smoother cell surfaces. The latter correlated with slower flocculation (due to reduced mannan-lectin interactions) [80].

Putrescine, spermidine and spermine, are cationic polyamines that affect growth and development by interacting with nucleic acids and membranes in fungi [81-83]. In zygomycetous fungi, polyamines were shown to be involved in DNA methylation and gene expression affecting morphogenesis [84]. In members of mucorales, spermine was not reported while spermidine levels in *M. bacilliformis* and *M. racemosus* yeast and hyphal cells were same while putrescine levels were 4-5 times higher in yeast cells than in the hyphal form cells [3, 83, 85-87].

Ornithine decarboxylase (ODC, EC 4.1.1.17) is the key enzyme that catalyses ornithine to putrescine conversion [88]. Increased levels of ODC and putrescine were observed during the Y-H transition in *M. rouxii*, *M. racemosus* and *M. bacilliformis* that suggested a role for polyamines in dimorphism [84]. The function of polyamines in growth and differentiation was demonstrated by the effect of the ODC inhibitors, 1,4- diamino-2-butanone (DAB) and, α-difluoromethylornithine (DFMO) [88]. The presence in *M. bacilliformis* yeast form mutants of low levels of ODC confirmed a role for polyamines in differentiation [89].

In *B.poitrasii* the levels of ODC and putrescine were higher in the hyphal form than the yeast form. The spermidine level was two fold higher in the hyphal form as compared to the yeast form. The decreased NADP-GDH activity in hyphal form could be correlated with the high levels of spermidine [90]. The exogenous addition of putrescine and spermidine enhanced the Y-H transition while decreased hyphal formation was observed in presence DAB and DFMO.

Chitin Metabolism

The enzyme involved in the synthesis of chitin, chitin synthases (CS), and in degradation, chitinase (EC 3.2.1.14) were reported to contribute significantly in fungal morphogenesis [91-93]. Bartinicki-Garcia [94] proposed the lytic-synthetic hypothesis of cell wall growth wherein the lytic enzymes plasticized the wall at the apex and also provided wall material for the synthesis of new cell walls. Both the enzymes, chitin synthase and chitinase showed form specific expression.

In *M. rouxii* the hyphal form associated CS was short-lived while yeast form associated CS was active for a longer period [95]. Various regulatory processes were suggested for the temporal and spatial control of CS activity in fungi [96]. For instance, CS activity in fungi was reported to be mostly zymogenic in nature and differential activation could be one of the factors which control morphogenesis in fungi. Ruiz-Herrera and Bartnicki-Garcia [95] reported that the yeast form CS was in a zymogenic form that required proteolytic activation and was stable for longer period. As CS is a membrane-bound enzyme the activity can also be affected by the membrane fluidity [97]. It was seen that in *B. poitrasii,* the CS activity was regulated by membrane stress as well as by proteolysis. However, the relationship between membrane stress and the activation by trypsinization was not clear. The native CS activity localized in *B. poitrasii* hyphal cell wall fractions was higher than that found with the yeast cell walls of parent or monomorphic mutant [97].

Chitinase complex, comprising of endochitinase (EC 3.2.1.1.14) and *N*-acetylglucosaminidase (EC 3.2.1.52) enzymes, was implicated in the assimilation of chitin, separation of dividing cells, branching of hyphae, autolysis, differentiation into spores, dimorphism and mycoparasitism [91, 98-99]. Rast *et al.* [98] reported multiple chitinase activities from the hyphal form in *M. rouxii*. In *M. mucedo*, the membrane bound chitinases contributed significantly to hyphal growth [100]. In *B. poitrasii* the endochitinase and *N*-acetylglucosaminidase activities increased during the yeast-hypha transition and the reverse was true during H-Y transition [101]. It was further reported that in *B. poitrasii* the *N*-acetylglucosaminidase activity increased up to 17 fold during Y-H transition, whereas endochitinase activity increased by 12 fold. Interestingly, one of the isozymes of *N*-acetylglucosaminidase present in the parent yeast and hyphal form was absent in yeast-monomorphic mutants, suggesting its possible role in the morphological transition [101].

The role of secondary messengers like calcium, calcium calmodulin (CaM) and cAMP are important in transducing the environmental signals that get translated into specific intracellular responses, important in metabolism, growth and differentiation [102]. CaM mediated the effect of calcium by phosphorylation and dephosphorylation of proteins that affected the growth and differentiation in dimorphic yeasts [103]. Apart from CaM, cAMP dependent protein kinases also reported to be involved in fungal dimorphism [104]. In *Mucor racemosus* aerobic Y-H change was halted by the addition of dbcAMP [105].The measurement of endogenous cyclic AMP contents showed that a four fold decrease in intracellular cAMP preceded the appearance of germ tubes. Similarly Paveto *et al.* [106] reported that in *M. rouxii* the Y-H transition was preceded by a decrease in cAMP level.

The exogenous addition of cAMP or its lipophilic derivative dbcAMP was associated with the yeast form, and halted Y-H transition in *B.poitrasii* [36, 67, 90]. In *B.poitrasii*, the phosphorylation/dephosphorylation of GDHs was reported to be correlated with the dimorphic behavior [36]. It was suggested that the relative proportion of active (phosphorylated) and less active (dephosphorylated) form-specific NADP-GDHs could be playing significant role in the morphology determination [36, 67]. Inhibitors of protein kinases (genistein, tyrosine protein kinase inhibitor; H-7, inhibitor of cyclic nucleotide-dependent protein kinase and protein kinase C), protein serine/threonine phosphatase, type 1A and 2A and 2B inhibitors (cantharidin and cyclosporine), and Ca^{++}. Channel blocker (verapamil), and Ca-CaM inhibitor (TFP, trifluoperazine), were studied to determine their effects on the Y-H transition in *B. poitrasii* [90, 97]. With the addition of any of the signal transduction inhibitors tested the Y-H transition was affected suggesting the significant role played by phosphorylation/dephosphorylation in *B. poitrasii* dimorphism [87, 94].

With the addition of cAMP dependent protein kinase inhibitor (H-7) the Y-H transition was found to be increased which could be attributed to a higher inhibition of phosphorylation of NADP-GDH of yeast form than the hyphal form in *B. poitrasii*. On the other hand, the presence of TFP inhibited phosphorylation of NADP-GDH of hyphal form and thus led to decrease in Y-H transition [67, 90].

To identify the cross talk between temperature and glucose, two dimorphism triggering factors in *B. poitrasii*, the high glucose and/or low temperature exposures were given to the yeast cells inocula, and the transition to the hyphal form was monitored [90]. The glucose effect on the enzyme activities (NAD-and NADP-GDH) and the morphological outcome was reversed by H-7 while TFP could reverse the effect of temperature on the morphological outcome.

Molecular Studies

Among the dimorphic *Mucor* species *M. rouxii*, *M. genevensis* and *M. bacilliformis* are homothallic while *M. racemosus* is heterothallic [3]; whereas *B. poitrasii* is homothallic [107]. The molecular mechanisms that control biochemical, signaling and morphological pathways can be understood with the help of morphological (monomorphic) mutants [22, 33, 72].

The dimorphic character of *M. circinelloides* was useful to understand the role of glucoamylase gene, important in starch degradation [108]. The extracellular enzyme secretion in the hyphal form was associated with the foraging nature for survival. The glucoamylase gene was repressed by glucose that commonly promoted the yeast form [108]. Glucoamylases were also reported from dimorphic zygomycete *Rhizopus oryzae* [109].

Dimorphism in *Mucor* is mainly affected by glucose that tranduces its effect through the secondary messenger cAMP, that in turn mediates its effects through protein kinase A (PKA). Binding of cAMP to the regulatory subunits (PKAR) resulted in the release of catalytic subunits (PKAC) subunits, triggering a kinase cascade affecting morphogenesis. Wolff *et al.* [110] cloned and characterised pkaR and pkaC encoding regulatory and catalytic subunits of the cAMP-dependent PKA of *M. circinelloides*. The expression levels of both pkaR and pkaC were significantly higher in anaerobically grown yeast cells than in the aerobically grown mycelium. However, there was 2-fold increase in expression of pkaR during the dimorphic shift. Further, overexpression of pkaR caused multi-branched colony phenotype [110]. The expression of both pkaR and pkaC was regulated in response to environmental perturbations affecting dimorphism, filamentation and branching.

The cAMP induced signal transduction pathway involves the ras proteins that are assigned a putative role in cell differentiation and proliferation [104]. Three RAS genes (*MRASI, MRAS2*, and *MRAS3*) from *M. racemosus* were cloned and characterized and found to be similar to other *ras* proteins [111]. High levels of MRASI transcripts were detected during Y-H morphogenesis while low levels were found during H-Y transition that implied its role in regulation of cAMP levels by suppression. MRAS3 was associated with the cAMP burst during spore germination while no transcript was detected for MRAS2 during transition or germination [111].

Cihlar and Sypherd [112] inserted the leucine biosynthetic genes of *M. racemosus* into *E. coli* plasmid pBR322 in another approach to understand the biosynthetic pathway. The recombinant plasmid complemented the leuB6 mutation in *E. coli*, that specified a defective form of isopropylmalate dehydrogenase. The effect was attributed to suppression by an unknown mechanism as *E. coli* is unable to assemble functional mRNA from the eukaryotic DNA template. Following the similar approach several highly conserved genes of *M. racemosus* such as genes for actin, α- and β-tubulin, and ras homologs were also cloned. [3, 111]. Using *S. cerevisiae* cDNA as probes, Linz and coworkers [113] cloned three distinct elongation factor lα (EF-lα) genes from *M. racemosus*. As EF-lα was attribute to have an essential role in eukaryotic translation it was suggested that its role in Y-H morphogenesis in *M. racemosus* could be explored with expression studies.

The cell wall components, chitin and chitosan are the main determinants of morphology. Relatively more chitosan to chitin ratios exist in the cell wall of zygomycetous fungi like *A. coerulea, M. rouxii, B. poitrasii, R. delemar, C. blackesleeana* and *M. isabelina* were reported [114]. In *B. poitrasii*, multiple chitin synthase genes were identified that were associated with the Y-H morphogenesis. Chitnis *et al.* [115] reported 8 chitin synthase genes with *BpCHS 1– 4* belonging to Classes I-III, *BpCHS 5-6* and *8* to Class IV and *BPCHS1-8*, genes *BpCHS7* of Class V in *B.poitrasii*. Among the 8 genes *BPCHS1-8*, 2 genes *BPCHS2* and *3* were specific to the hyphal form. Chitin synthase Class II enzymes were reported to synthezise septal chitin in *S. cerevisiae* cells and *C. albicans* whereas it had additional role in lateral wall synthesis in the later [116-118]. The Class II enzyme of *A. nidulans* was suggested to be involved in conidium formation and hyphal wall synthesis [119]. Class V chitin synthase of *A. fumigatus* synthesized 25% of hyphal wall chitin and was also reported to be involved in sporulation [120]. Presence of multiple chitin synthase genes in *B. poitrasii* suggested that such fungi may have an intermediate position in comparison to filamentous fungi and more advanced yeasts. Similar studies are being carried out for the glutamate dehydrogenase genes from *B. poitrasii*. The form specific expression of NADP-GDH gene is being explored in connection with the Y-H transition.

According to the model proposed by Davis and Bartnicki-Garcia for the biosynthesis of chitin and chitosan, nascent chitin chains are modified by chitin deacetylases (CDAs) during their synthesis by the enzyme chitin synthase [3]. The synergistic action of CS – *CDA* resulted in chitin - chitosan containing fibrils that crystallized to form the main structural mesh of the cell walls. *CDA* could also be involved in deacetylation of chitin oligosaccharides during autolysis after action of endo-chitinases on the cell wall. The first *CDA* gene isolated, characterized and sequenced was from *M. rouxii* [121]. In *S. cerevisiae* genes *CDA1* and *CDA2* were identified and sequenced by homology comparison to *M. rouxii CDA1* gene from the cosmid library constructed for the yeast [122, 123]. *CDA* genes from *Gongronella butleri*, and *R. nigricans* were also cloned and characterized [124, 125].

Mathematical Models For Wall Growth in Zygomycetous Fungi

The manner in which fungi attain their characteristic shape has generated interest among theoretical and experimental biologists [126, 127]. A number of mathematical models describing the pattern of surface growth leading to generation of a tubular shape were developed to explain the hyphal tip growth. These mathematical models were based on equations formulated from a number of artificial co-ordinates and reference points for a virtual geometric understanding of hyphal growth. Trinci and Saunders [128] revised the prevailing hemispherical tip model with the modified cotangent function to take into consideration the "half –ellipsoidal"shape of the growing tip. Bartinicki-Garcia and Lippman [129] studied the hyphal tip growth in *M. rouxii* by tracing the radiolabelled *N*-acetylglucosamine, a chitin/chitosan precursor. A pattern of deposition confirmed the descending gradient of wall synthesis from the centre of the hyphal tip [129, 130]. Growth and differentiation during the fungal life cycle usually involves three types of wall deposition patterns. The growth of the germ tube during spore germination showed a concentration of the precursors at the hyphal tip that was in a polarized regulated manner for hyphal elongation [129]. In the spherical spore, a regulated non-polarized deposition pattern is prevalent, and occasionally deregulated polarized deposition pattern triggers irregular growth. The dimorphic fungi display an equilibrium between spherical growth (budding) and polarized (hyphal or pseudohyphal tip elongation). Prosser and Trinci [131] proposed that vesicles containing wall precursors and/or enzymes required for wall synthesis travel to the hyphal tip and fuse with the plasma membrane for tip elongation. These predictions were experimentally compared with growth kinetics of *Apergillus nidulans* and *Geotrichum candidum*. Bartnicki-Garcia *et al.* [132] proposed an empirical model where cell wall construction was carried out from vesicles distributed from a sub-apical 'Spitzenkörper' assembly. Spitzenkörper or vesicle supply centre (VSC) represented the assembly of vesicles that arrive from the sub-apical regions of the hypha. According to the model, the ovoid/ellipsoid shape typical of the yeast form were generated by invoking slow displacement of VSC and high rate of vesicle generation. The spherical buds typical of yeast resulted if there was no net displacement of the VSC. Clearly the hyphal or yeast form could be attained by the VSC and the rate at which the vesicles were distributed from it.

Recently, the dimorphic fungi *C. boidinii* and *Yarrowia lipolytica* were used to understand the hyphal growth and the model fitted the hypothesis when colonies were regarded as well-separated individual cells whose proliferation was limited by diffusive transport of glucose [133]. In case of colonies, the simulations and experimental results were in agreement for hyphal development when biomass recycling and cannibalizing growth were incorporated into the model. The use of zygomycetous fungi that respond to glucose and temperature as environmental stimuli for hyphal growth may be useful to understand dimorphic growth.

Fungal Evolution

Among the fungi, the ascomycetes and basidiomycetes were well studied at the molecular phylogenetic level in comparison to the zygomycetes [134, 135]. Voigt and Wostermeyer [136] carried out a combined phylogenetic analysis, using the genes for actin (act) and translation elongation factor (EF-1α), to study the evolutionary history of the zygomycetes. Both genes were highly conserved and exhibited low rates of amino acid substitutions and were single or low copy number genes in all taxa examined to date. A polyphyletic origin of the zygomycetes was suggested as the Entomophthorales, Glomales and Endogonales, Kickxellales, Mortierellales and Mucorales formed different evolutionary lineages whose evolution was linked with that of the chytridiomycetes [136, 137]. O'Donnell *et al.* [138] placed the dimorphic *B. poitrasii* of family Mycotyphaceae and *C. recurvatus* of family Thamnidiaceae together on basis of the analyses using 18S, 28S rRNA gene subunits and EF-1 α.

In the evolution of fungi the yeasts are considered as evolved and the significance of yeasts in the phylogeny of fungi has been reviewed extensively [139]. During the evolution process, yeast possibly evolved from filamentous fungi by a series of changes in mechanisms of cell wall growth and deposition pattern. The induction of yeast like vegetative growth in the zygomycetous dimorphic fungi such as *Mucor* and *Rhizopus* was reported to be significant in the phylogeny of Eumycota [140]. The germination of the sporangiospores and the zygospores of *B. poitrasii* into the yeast form was reported to be important in understanding the evolutionary position of dimorphic zygomycetes among the more evolved yeasts and the filamentous fungi [23].

Morphology is determined by the cell wall composition and the difference in the polysaccharide components were useful in tracing the fungal phylogeny [139, 141-143]. In this context it was suggested that dimorphic fungi with varying mannan contents in the yeast and hyphal cells may form an intermediate group in the evolution of fungi [23].

The enzyme from the nitrogen assimilation pathway, glutamate dehydrogenase, that contributes to cell wall chitin contents has evolutionary significance in the fungal realm. LeJohn [68] compared the enzyme structure, regulation and the biochemical pathways to speculate the convergent evolution of NAD-GDHs of class I(unregulated) and Class II(divalent ion regulation) in the Mucorales.

The common occurrence of chitin among the fungi, insects, and arthropods suggested an early evolutionary appearance of chitin synthase. The chitin synthases probably shared an ancestor with different β- glycosyl-transferases, as suggested by their common structural features [144]. Bowen *et al.* [145] grouped chitin synthases into class I to VI that were later categorized by Ruiz-Herrera and Ortiz-Castellanos [146] into division 1 containing CHS classes I-III and division 2 representing classes IV-V. Among fungi ascomycetes contained all CHS classes from division 1, basidiomycetes lack one of them (class I) while zygomycetes, and chytridiomycetes possess only one (class II), whereas primitive Microsporidia contain none [147]. *Rhizopus oligosporus* Class II genes (*CHS1* and *CHS2*) were reported to function mainly in the growing hyphal form. The dimorphic fungus *B. poitrasii* was reported to have 8 chitin synthase genes with *BpCHS 1 – 4* belonging to classes I-III, *BpCHS 5-6* and *8* of Class IV and *BpCHS 7* of Class V [115]. Among CS genes reported for *B. poitrasii* 2 genes *BpCHS 2* and *BpCHS 3* appeared to be hyphal form specific. Chitin synthase Class II enzymes were reported for septal chitin synthesis, conidium formation and hyphal wall synthesis in ascomycetes [119]. Presence of multiple chitin synthase genes in the *B. poitrasii* a dimorphic fungus suggested that such fungi may have an intermediate position in comparison to filamentous fungi and more advanced yeasts. Requelme and Bartinicki-Garcia [148] analyzed the phylogentic relationship in higher fungi on the basis of chitin synthase genes. They further suggested that dimorphic fungi such as *Benjaminiella*, *Candida*, *Histoplasma* possessed multiple chitin synthase genes that could be attributed to the composition of the cell wall and its biological function.

Chitosan is mainly the characteristic component of the zygomycetous cell walls. In the evolutionary context, Ruiz-Herrera and Ortiz-Castellanos [146] reported the presence of CDAs among the primitive fungi (chytridiomycete and microsporidian species). The mucoraceous fungi analyzed showed the presence of multiple CDAs that were associated with the chitosan contents of the zygomycetous cell walls. Futhermore, multiple *CDA* were suggested to have arisen due to repeated processes of gene duplication along evolution in this group [146].

Biotechnological Applications of Dimorphic Zygomycetes

In general, filamentous fungi secrete large amounts of extracellular proteins from growing hyphal tips [149, 150]. However, unlike unicellular yeast, in submerged fermentation they form either aggregates or pellets and sometimes grow in dispersed mycelial form. The dimorphic *M. circinelloides* grew rapidly as multipolar yeasts thus producing biomass efficiently in a homogenous culture. After sufficient biomass was obtained, the Y-H transition was induced that provided optimal conditions for protein secretion. Control of biochemical and genetic factors affecting the morphology will be the approaches useful for optimization and subsequent production of recombinant proteins by fermentation. For instance, morphological shift from

Y-H growth can be achieved by changing anaerobic environment to aerobic condition. Exogenous addition of cAMP resulted in the constitutive yeast growth in *M. circinelloides* [3]. cAMP is used by cAMP dependant protein kinase A (PKA) which consists of two regulatory subunits (PKAR) that bind to and inhibit the activity of two catalytic subunits (PKAC). Expression levels of *pkaR* and *pkaC* were found to be higher in anaerobic yeast culture whereas overexpression of *pkaR* resulted hyper-branching morphologies [151]. Thus, in *M. circinelloides* , thus the Y-H change was used for development of an expression system for the recombinant protein [42, 110, 151]. Thus Y-H transition of *Mucor* species can be exploited for the production of heterologous proteins such as glucose oxidase, glyceraldehyde-3-phosphate dehydrogenase and others. For instance, the production of *A. niger* glucose oxidase and native amylase was carried out by the recombinant *M. circinelloides* strain by altering the dimorphic growth. The shift in anaerobic to aerobic conditions was used to promote the transition to the hyphal form that produced glucose oxidase and amylase [152].

High alkaline protease activity was reported in the members of order entomophthorales, *C. coronatus* [153] and *Basidiobolus* [30]. Ingale *et al.* [30] reported that in *Basidiobolus* species, a number of inorganic salts triggered the darmform morphology in the submerged fermentation. This transition was associated with the significant increase in the levels of alkaline protease activity while the pelleted growth was associated with less extracellular protease activity. This study could be useful for developing a technology for the production of alkaline protease based on submerged fermentation.

The conversion of polysaccharides like starch and cellulose and the disaccharide lactose to glucose and further to ethanol can be made by the use of the dimorphic transition in *B. poitrasii* in a step-wise manner [44]. In the presence of starch, cellulose and even lactose *B. poitrasii* grew in the hyphal form and breakdown of these carbon sources resulted in glucose formation. As glucose is one of the dimorphism triggers in *B. poitrasii* the subsequently formed yeast cells converted sugar into ethanol [22]. Such a strategy can be used by identifying other efficient dimorphic organisms for commercial exploitation.

The cell walls of *B. poitrasii* hyphal form contained 1.7 times more hexosamine than the yeast form [38]. However, the chitosan/chitin ratio was found to be more in the yeast form than the hyphal form. With glucose and complex nitrogen in the submerged fermentation the yeast biomass can be produced in large quantities and thus suitable for growth in fermentor for recovery of chitosan on large scale. Alternately, the hyphal form can be used for chitosan production using a *CDA* treatment for the conversion of chitin to chitosan [114, 154].

Biodiesel, consisting of fatty acid methyl esters (FAMEs), is usually obtained from plant oils. Among the fungi, oleaginous zygomycetes species, such as *M. isabelina* and *C. echinulata*, which may accumulate up to 86 and 57% of lipids in dry biomass, represented potential biodiesel sources [155]. *M. circinelloides* grown with high glucose under aerobic conditions produced both yeast and hyphal cells [7]. Yeast form cells contained more polar lipids and free fatty acids and less principal reserve lipids (triacylglycerides) than hyphal cells. Vicente *et al.* [156] reported that *M. circinelloides* hyphal biomass from submerged cultures was a suitable feedstock for biodiesel production. It would be interesting to trigger Y-H transition for more hyphal mass to obtain biodiesel.

Perspective

The morphological outcome in fungi is dependent on the physiological make up of the cell. Among the zygomycetous fungi, *Mucor* is one of the well studied fungus for a variety of purposes including healthcare, biotechnological applications and developmental aspects [3]. Other than *Mucor* species, a candidate which has been included in the discussion is *B. poitrasii* which has also been studied extensively [42, 70]. One of the emerging areas is nanobio-technology and more uniform biologically synthesized silver and gold nanoparticles were obtained using *Rhizopus stolonifer* [156]. The preliminary experiments with *B. poitrasii* yeast and hyphal cells showed differential synthesis of gold nanoparticles. It will be interesting to explore dimorphic fungi for the nanomaterial synthesis. Another interesting area, is gene silencing by RNA interference (RNAi) which is in its initial stages in fungal systems [157]. The utility of the method was explored by studying the developmental defects with dicer disruptants in dimorphic *M. circinelloides*. The dicer disruptants appeared to have slight morphological abnormalities in vegetative mycelia [158]. The dimorphic transitions in zygomycetes can be studied with gene silencing. RNAi can be a good tool for gene function analysis in fungi. The understanding of biochemical and molecular mechanism of dimorphism using novel approaches indeed would give the insight for metabolic engineering for a variety of applications.

ACKNOWLEDGEMENTS

Authors are grateful to Department of Science and Technology, New Delhi for the support.

REFERENCES

[1] White MM, James T Y, O'Donnell K, Cafaro MJ, Tanabe Y, Sugiyama J. Phylogeny of the Zygomycota based on nuclear ribosomal sequence data. Mycologia 2006; 98: 872-884.

[2] Kirk PM, Cannon PF, David JC, Stalpers J. Ainsworth and Bisby's Dictionary of the Fungi. 9th ed. CAB International, Wallingford, UK. 2001.

[3] Orlowski M. *Mucor* dimorphism. Microbiol Rev 1991; 55: 234-258.

[4] Bartnicki-Garcia S, Nickerson WJ. Nutrition, growth and morphogenesis of *Mucor rouxii*. J Bacteriol 1962; 84: 841-858.

[5] Omoifo CO. Sporangiospore-to-yeast conversion: Model for morphogenesis. Afric J Biotechnol 2009; 8: 3854-3863.

[6] Mysyakina I, Funtikova N. Lipid composition of the yeast like and mycelial *Mucor hiemalis* cells grown in the presence of 4-chloroaniline Microbiol 2000; 69: 670-675.

[7] Mysyakina I, Funtikova N The role of sterols in morphogenetic processes and dimorphism in fungi. Microbiol 2007; 76: 1-13.

[8] Sypherd PS, Borgia PT, Paznokas JL. The biochemistry of morphogenesis in the fungus Mucor. Adv Microb Physiol 1978; 8: 67-104.

[9] Gow NAR. Yeast-hyphal dimorphism. In: Gow NAR, Gadd GM, Eds. Growing fungus. Chapman and Hall, London. 1995; pp 403-422.

[10] Terenzi F, Storck R. Stimulation by phenethyl alcohol of aerobic fermentation in *Mucor rouxii*. Biochem Biophy Res Com 1968; 30: 447-452.

[11] Serrano T, Da Silva L, Roseiro JC. Ethanol-induced dimorphism and lipid composition changes in *Mucor fragilis* CCMI 142. Lett Appl Microbiol 2001; 33: 89-93.

[12] Benny GL, Kirk PM, Samson RA. Observation on Thamnidiaceae (Mucorales). III. Mycotyphaceae fam. Nov. and a re-evaluation of *Mycotypha* sensu Benny and Benjamin illustrated by two new species. Mycotaxon 1985; 22 : 119-48.

[13] Price JS, Storck R, Gleason FH. Dimorphism of *Cokeromyces poitrasii* and *Mycotypha microspora*. Mycologia 1973; 65: 1274-1283.

[14] Hall MJ, Kolankaya N. The physiology of mold yeast dimorphism in the genus *Mycotypha* (Mucorales). J Gen Microbiol 1974; 82: 25-34.

[15] Schulz BE, Kraepelin G,Hinkelmann W. Factors affecting dimorphism in *Mycotypha* (Mucorales): a correlation with the fermentation/respiration equilibrium. J Gen Microbiol 1974; 82: 1-13.

[16] Delgado AE, Urdaneta LM, Piñeiro Chávez AJ. *Mycotypha indica* P.M. Kirk & Benny, in turkey dung, a new record for Venezuela. Multiciencias 2007; 7: 176-180.

[17] Ryan LJ, Ferrieri P, Powell R, Zeki S, Pambuccian S. Fatal *Cokeromyces recurvatus* pneumonia: report of a case highlighting the potential for histopathologic missdiagnosis as *Coccidoides*. Int J Surg Pathol 2009; DOI 10.1177/1066896908330483.

[18] Kemna ME, Neri RC, Ali R, Salkin I. *Cokeromyces recurvatus*, a mucoraceous zygomycete rarely isolated in clinical laboratories. J Clin Microbiol 1994; 32: 843-845.

[19] Nielsen C, Sutton DA, Matise I, Kirchhof N, Libal MC. Isolation of *Cokeromyces recurvatus*, initially misidentified as *Coccidioides immitis*, from peritoneal fluid in a cat with jejunal perforation. J Vet Diagnostic Invest 2005; 17: 372-378.

[20] Cole GT, SekiyaT, Kasai R, Nozawa Y. Morphogenesis and wall chemistry of the yeast, "intermediate", and hyphal phases of the dimorphic fungus, *Mycotypha poitrasii*. Can J Microbiol 1979; 26: 37-49.

[21] Kirk PM. A new species of *Benjaminiella* (Mucorales : Mycotyphaceae) Mycotaxon 1989; 35: 121-125.

[22] Khale A, Srinivasan MC, Deshmukh SS, Deshpande MV. Dimorphism of *Benjaminiella poitrasii*: isolation and biochemical studies of morphological mutants. Ant van Leeuwenhoek 1990; 57: 37-41.

[23] Ghormade V, Deshpande MV. Fungal spore germination into yeast or mycelium: Possible implications of dimorphism in evolution and human pathogenesis. Naturwissenschaften 2000; 87: 236-240.

[24] Ghormade V, Sainkar S, Joshi C, Doiphode N, Deshpande M. Dimorphism in *Benjaminiella poitrasii*: Light, fluorescence and scanning electron microscopy studies of the vegetative and reproductive forms with special reference to the glutamate dehydrogenase, a novel fungicidal target. J Mycol Pl Pathol 2005; 35: 1-11.

[25] Karmakar M, Ray RR. Extracellular endoglucanase production by *Rhizopus oryzae* in solid and liquid state fermentation of agro wastes. Asian J Biotechnol 2010; 2: 27-36.

[26] Ribes JA, Vanover-Sams CL, Baker DJ . Zygomycetes in human disease. Clinical Microbiol Rev 2000; 13: 236–301.

[27] Santos MJS, de Oliveira PC, Trufem S. Morphological observations on *Absidia corymbifera* and *Absidia blakesleeana* strains preserved under mineral oil. Mycoses 2003; 46: 402-406.

[28] Hejtmánek M, Kodousek R, Raclavský V. The fluorescence brightener Rylux BSU induces dimorphism in *Basidiobolus ranarum*. Folia Microbiol 1993; 38: 395-398.

[29] Haplova J, Farkas V, Hejtmánek M, Kodousek R, Malinsky J. Effect of the new fluorescent brightener Rylux BSU on morphology and biosynthesis of cell walls in *Saccharomyces cerevisiae*. Arch Microbiol 1994; 161: 340-344.

[30] Ingale SS, Rele MV, Srinivasan MC. Alkaline protease production of *Basidiobolus* (N.C.L. 97.1.1): effect of 'darmform' morphogenesis and cultural conditions on enzyme production and preliminary enzyme characterization. World J. Microbiol Biotechnol. 2002; 18: 403-408.

[31] Humber R. Synopsis of a revised classification for the Entomopthorales (Zygomycotina) Mycotaxon 1989; 34:441-460.

[32] Tonka T, Weiser J, Weiser J. Budding: A new stage in the development of *Chytridiopsis typographi* (Zygomycetes: Microsporidia) J Invertebr Pathol 2010; 104: 17-22.

[33] Ruiz-Herrera J, Ruiz A, Lopez-Romero E. Isolation and biochemical analysis of *Mucor bacilliformis* monomorphic mutants. J Bacteriol 1983; 156: 264-272.

[34] Storck R, Morrill RC.Respiratory deficient, yeast-like mutant of Mucor. Biochem Genet 1971; 5:467-479.

[35] Khale A, Srinivasan MC, Deshpande MV. Significance of NADP/NAD-glutamate dehydrogenase ratio in the dimorphic behavior of *Benjaminiella poitrasii* and its morphological mutants. J Bacteriol 1992; 174: 3723-3728.

[36] Khale-Kumar A, Deshpande MV. Possible involvement of cyclic adenosine 3', 5'- monophosphate in the regulation of NADP-/NAD-glutamate dehydrogenase ratio and in yeastmycelium transition of *Benjaminiella poitrasii*. J Bacteriol 1993; 175: 6052-6055.

[37] Bartnicki-Garcia S, Nickerson WJ. Induction of yeast-like development in *Mucor* by carbon dioxide. J Bacteriol 1962; 84: 829-840.

[38] Khale A, Deshpande MV. Dimorphism in *Benjaminiella poitrasii*: cell wall chemistry of parent and two stable yeast mutants. Ant Van Leeuwenhoek 1992; 62: 299-307.

[39] Chitnis MV, Deshpande MV. Isolation and regeneration of protoplasts from the yeast and mycelial form of the dimorphic zygomycete *Benjaminiella poitrasii*: Role of chitin metabolism for morphogenesis during regeneration. Microbiol Res 2002; 157: 29-37.

[40] Chitnis MV. A PhD thesis submitted to Pune University, Pune India, 2001.

[41] Chitnis MV, Deshpande MV. Fungal protoplast technology. In: Deshmukh SK, Rai MK, Eds. Biodiversity of Fungi, their Role in Human Life, Oxford & IBH Publishing Co. Pvt.Ltd., New Delhi, 2005; pp 439-454.

[42] Doiphode N, Joshi C, Ghormade V, Deshpande MV. The biotechnological applications of dimorphic yeasts. In: Satyanarayana T, Kunze G, Eds. Yeast Biotechnology: Diversity and Applications, Springer Science + Business Media B.V. 2009; pp 635-650.

[43] Wöstermeyer A, Wöstermeyer J. Fungal protoplasts: relics or modern objects of molecular research? Microbiol Res 1998; 153: 97-104.

[44] Reyes E, Novaes-Ledieu M, Garcia-Mendosa C. Preparation of mycelia and yeast-like spheroplasts from *Mucor rouxii* by a lytic enzyme preparation form *Penicillium islandicum*. Curr Microbiol 1983; 9: 301-304.

[45] Ramirez-Leon IF, Ruiz-Herrera J. Hydrolysis of walls and formation of sphaeroplasts in *Mucor rouxii*. J Gen Microbiol 1972; 72: 281-290.

[46] Lasker BA, Borgia PT. High frequency heterokaryon formation by *Mucor racemosus* J Bacteriol 1980; 141: 565-569.

[47] Matsunobu T, Hiruta O, Nakagawa K, Murakami H, Miyadoh S, Uotani K, Takabe H, Satoh A. A novel chitosanase from *Bacillus pumilus* BN262, properties, production and applications. Sci Rep 1997; 35: 28-50

[48]   Gabriel M. Formation and regeneration of protoplasts in the mold *Rhizopus nigricans*. Folia Microbiol 1968; 13: 231-234.

[49]   Wöstemeyer J, Brockhausen-Rohdemann E. Inter-mating type protoplast fusion in the zygomycete *Absidia* glauca. Curr Genet 1987; 12 :435-441.

[50]   Binding H, Weber HJ. The isolation, regeneration and fusion of Phycomyces protoplasts. Mol Gen Genet 1974; 135: 273-276.

[51]   Peberdy JF. Fungal cell wall: a review. In: Kuhn PJ, Trinci APJ, Jung MJ, Goosey MW, Cooping LG, Eds. Biochemistry of Cell Walls and Membranes in Fungi. Springer-Verlag, Berlin, 1990; pp 5-30.

[52]   Wessels JGH, Mol PC, Sietsma JH, Vermeulen CA. Wall structure, wall growth and fungal cell morphogenesis. In: Kuhn P J, Trinci APJ, Jung MJ, Goosey MW, Copping LG, Eds. Biochemisty of Cell Walls and Membranes in Fungi. Springer-Verlag, Berlin 1989; pp 81-95.

[53]   Gooday GW. Cell walls In: Gow NAR, Gadd GM, Eds. Growing fungus. Chapman and Hall, London 1995; pp 43-59.

[54]   Valdivieso MH, Dura n A, Roncero C. Chitin biosynthesis and morphogenetic processes. In : Brambl R, Marzluf GA, Eds. The Mycota, vol. III. Springer, Berlin, Heidelberg, New York 2004; pp 275-290.

[55]   Bartnicki-Garcia S. Role of chitosomes in the synthesis of fungal cell walls. In: Schlessinger D, Ed. Microbiology. American Society for Microbiology, Washington DC, 1981; pp 238-241.

[56]   Ruiz-Herrera J. Fungal Cell Wall. Structure, Synthesis and Assembly. CRC Press. Boca Raton 1992.

[57]   Bartnicki-Garcia S. Cell wall chemistry. Ann Rev Microbiol 1968; 22: 87-108.

[58]   Gooday GW, Trinci APJ. Wall structure and biosynthesis in fungi. In: Gooday GW, Lloyd D, Trinci APJ, Eds. The Eukaryotic Microbial Cell. Cambridge University Press, Cambridge. Vol 30, 1980; pp. 207-251.

[59]   Domek DB, Borgia PT. Changes in the rate of chitin-plus-chitosan synthesis accompany morphogenesis of *Mucor racemosus*. J Bacteriol 1981; 146: 945-951.

[60]   Bartnicki-Garcia S, Nickerson WJ. Isolation, composition, and structure of cell walls of the filamentous and yeast-like forms of *Mucor rouxii* Biochim Biophys Acta 1962; 58: 102-119.

[61]   Bartnicki-Garcia S. Mold-yeast dimorphism of *Mucor*. Bacteriol Rev 1963; 27: 293-304.

[62]   Barrera CR, Corral J. Effect of hexoses on the levels of pyruvate decarboxylase in *Mucor rouxii*. J Bacteriol 1980; 142: 1029-1031.

[63]   Paznokas JL, Sypherd PS. Pyruvate kinase isozymes of *Mucor racemosus*: control of synthesis by glucose. J Bacteriol 1977; 130: 661-666.

[64]   Inderlied CB, Sypherd PS. Glucose metabolism and dimorphism in *Mucor*. J Bacteriol 1978; 133: 1282-86.

[65]   Rogers PJ, Gleason FH. Metabolism of *Cokeromyces poitrasii* grown in glucose-limited continuous culture at controlled oxygen concentrations. Mycologia 1974; 66: 921-925.

[66]   Rangel-Porras RA, Meza-Carmen V,. Martinez-Cadena G,. Torres-Guzmán JC, González-Hernández GA, Arnau J,. Gutiérrez-Corona JF. Molecular analysis of an NAD-dependent alcohol dehydrogenase from the zygomycete *Mucor circinelloides*, Mol Gen Genomics 2005; 274: 354-363.

[67]   Amin A, Joshi M, Deshpande MV. Morphology-associated expression of NADP-dependent glutamate dehydrogenases during yeast-mycelium transition of a dimorphic fungus *Benjaminiella poitrasii*. Ant van Leeuwenhoek 2004; 85: 327-334.

[68]   LéJohn HB. Enzyme regulation, lysine pathways and cell wall structures as indicators of major lines of evolution in fungi. Nature 1971; 231: 164 -168.

[69]   Marzluf GA. Genetic regulation of nitrogen metabolism in fungi. Mol Biol Rev 1997; 61: 17-32.

[70]   Deshpande MV. Biochemical basis of fungal differentiation. In: Varma A, Ed. Microbes: For Health, Wealth and Sustainable Environment. Malhotra Publishing House, New Delhi, 1998; pp 241-252.

[71]   Avendano A, DeLuna A, Olivera H, Valenzuela L, Gonzalez A. GDH3 encodes a glutamate dehydrogenase isozyme previously unrecognized route for glutamate biosynthesis in *Saccharomyces cerevisiae*. J Bacteriol 1997; 179: 5594-5597.

[72]   Peters J, Sypherd PS. Morphology-associated expression of NAD-dependent glutamate dehydrogenase in *Mucor racemosus*. J Bacteriol 1979; 137: 1137-1139.

[73]   Green J, Large PJ. Regulation of the key enzymes of methylated amine metabolism in *Candida boidii*. J Gen Microbiol 1984; 130: 1947-1959.

[74]   Fergusson AR, Sims AP. The regulation of glutamine metabolism in *Candida utilis*: the inactivation of glutamine synthetase 1974; 80: 173-185.

[75]   Holmes AR, Collings A, farnden KJF, Shepherd MG. Ammonium assimilation in *Candida albicans* and other yeasts: evidence for the activity of glutamate synthetase. J Gen Microbiol 1989; 135: 1423-1430.

[76]   Chattaway FW, Bishop R, Holmes MR, Odds FC. Enzyme activities associated with carbohydrate synthesis and breakdown in the yeast and mycelial forms of Candida albicans. J Gen Microbiol 1973; 75: 97-109.

[77]   Wills EA, Redinho MR, Perfect JR, Poeta MD. New potential targets for antifungal development. Emerging Therapeut Targ 2000; 4: 1-32.

[78]   Boles E, Lehnert W, Zimmermann W. The role of the NAD - dependent glutamate dehydrogenase in restoring growth on glucose of a *Saccharomyces cerevisiae* phosphoglucose isomerase mutant. Eur J Biochem 1993; 217: 469-477.

[79]   Osmond BC, Specht CA, Robbins PW Chitin synthase III: synthetic lethal mutants and "stress related" chitin synthesis that bypasses the CSD3/CHS6 localization pathway. Biochem 1999; 96: 11206-11210.

[80]   Joshi C, Ghormade V, Kunde P. Kulkarni P, Mamgain H, Bhat S, Paknikar KM. Flocculation of dimorphic yeast *Benjaminiella poitrasii* is altered by modulation of NAD-glutamate dehydrogenase. Bioresource Technol 2010; 101: 1393-1395.

[81]   Shapira R, Altman A, Henis Y, Chet I. Polyamines and ornithine deacrboxylase activity during growth and differentiation in *Sclerotium rolfsii.* J Gen Microbiol 1989; 135: 1361-1367.

[82]   Smith TA, Barker JHA, Jung M. Growth inhibition of *Botrytis cinerea* by compounds interfering with polyamine metabolism J Gen Microbiol 1990; 136: 985-992.

[83]   Marshall M, Russo G, Van Etten J, Nickerson K. Polyamines in dimorphic fungi. Curr Microbiol 1979; 2: 187-190.

[84]   Ruiz- Herrera J. Polyamines, DNA methylation and fungal differentiation. Crit Rev Microbiol 1994; 20: 143-150.

[85]   Garcia R, Hiatt W, Peters J, Sypherd PS. Adenosylmethionine levels and protein methylation during morphogenesis of *Mucor racemosus.* J Bacteriol 1980; 142: 196-201.

[86]   Calvo-Mendez C, Ruiz-Herrera J. Regulation of S-Adenosylmethionine decarboxylase during the germination of sporangiospores of *Mucor rouxii.* J Gen Microbiol 1991, 137: 307-314.

[87]   Nickerson JW, Dunkle LD, Van Etten JL. Absence of spermine in filamentous fungi. J Bacteriol 1977; 129: 173-176.

[88]   Martinez-Pacheco M, Ruiz-Herrera J. Differential compartmentalization of ornithine deacarboxylase in cells of *Mucor rouxii.* J Gen Microbiol 1993; 139: 1387-1394.

[89]   Ruiz-Herrera J. The role of polyamines in fungal cell differentiation. Arch Med Res 1993; 24: 263-265.

[90]   Ghormade V, Joshi C, Deshpande MV. Regulation by polyamines: A possible model for signal transduction pathway leading to dimorphism in *Benjaminiella poitrasii* J Mycol Pl Pathol 2005; 35: 442-450.

[91]   Patil RS, Ghormade V, Deshpande MV. Chitinolytic enzymes: an exploration. Enzyme Microbial Technol 2000; 26: 473-483.

[92]   Sahai AS, Manocha MS. Chitinases of fungi and plants: their involvement in morphogenesis and host-parasite interaction. FEMS Microbiol Rev 1993; 11: 317-338.

[93]   Adams DJ. Fungal cell wall chitinases and glucanases. Microbiology 2004; 150: 2029-2035.

[94]   Bartinicki-Garcia S. Fundamental aspects of hyphal morphogenesis, In: Ashworth JM, Smith JE, Eds. Microbial Differentiation Cambridge University Ptress, Cambridge 1973; pp 245-267.

[95]   Ruiz-Herrera J, Bartinicki-Garcia S. Proteolytic activation and inactivation of chitin synthetase from *Mucor rouxii.* J Gen Microbiol 1976; 97: 241-249.

[96]   Ramirez-Ramirez N, Gutierrez-Corona F, Lopez-Romero E. Nikkomycin-resistant mutants of *Mucor rouxii:* physiological and biochemical properties. Ant van Leeuwennoek 1993; 64: 27-33.

[97]   Deshpande MV, O'Donnell R, Gooday GW. Regulation of chitin synthase activity in the dimorphic fungus *Benjaminiella poitrasii* by external osmotic pressure. FEMS Microbiol Lett 1997; 152: 327-332.

[98]   Rast DM, Horsch M, Furter R, Gooday GW. Complex chitinolytic system in exponentially growing mycelium of *Mucor rouxii:* properties and function. J Gen Microbiol 1991; 137: 2797-2810.

[99]   Pedraza-Reyes M, Lopez-Romero E. Detection of nine chitinase species in germinating cells of *Mucor rouxii.* Curr Microbiol 1991; 22: 43-46.

[100] Humphreys AM, Gooday GW. Properties of a chitinase activities from *Mucor mucedo:* evidence for a membrane bound zymogenic form. J Gen Microbiol 1984; 130: 1359-1366.

[101] Ghormade V, Lachke SA, Deshpande MV. Dimorphism in *Benjaminiella poitrasii:* Involvement of intracellular endo-chitinase and N-acetylglucosaminidase activities in the yeast-mycelium transition. Folia Microbiol 2000; 45: 231-238.

[102] Gadd GM. Signal transduction in fungi. In: Gow NAR, Gadd GM, Eds. Growing fungus. Chapman and Hall, London 1995; pp 403-422.

[103] Anraku Y, Ohya Y, Iida H. Cell cycle control by calcium and calmodulin in *Saccahromyces cerevisiae.* Biochem Biophys Acta 1991; 1093: 169-177.

[104] Gancedo JM, Mazon MJ, Eraso P. Biological roles for cyclic AMP: similarities and differences between organisms. Trends Biochem Sci 1985; 10: 210-212.

[105] Larsen AD, Sypherd PS. Cyclic adenosine 3', 5'-monophosphate and morphogenesis in *Mucor racemosus*. J Bacteriol 1974; 117: 432-438.

[106] Paveto C, Epstein A , Passeron S. Studies on cyclic adenosine 3',5'-monophosphate levels, adenylate cyclase and phosphodiesterase activities in the dimorphic fungus *Mucor rouxii*. Arch Biochem Biophys 1975; 169: 449-457.

[107] Ghormade V, Shastry P, Chiplunkar J, Deshpande MV. Determination of ploidy of a dimorphic zygomycete *Benjaminiella poitrasii* and the occurrence of meiotic division during zygospore germination. Agric Technol 2005, 1: 97-112.

[108] Houghton-Larsen J, Pedersen PA. Cloning and characterization of a glucoamylase gene (GlaM) from the dimorphic zygomycte *Mucor circinelloides*. Appl Microbiol Biotechnol 2003; 62: 210-217.

[109] Ashikari T, Nakamura N, Tanaka Y, Kiuchi N, ShiBano Y, Tanaka T, Amachi T, Yoshizumi H. *Rhizopus* raw-starch degrading glucoamylase: its cloning and expression in yeast. Agric Biol Chem 1986; 50: 957-964.

[110] Wolff A, Appel K, Petersen J, Poulsen U, Arnau J, Jacobsen M. Recombinant dimorphic fungal cell. 2002. WO/2002/070721

[111] Casale WL, McConnell DG, Wang SY, Lee YJ, Linz JE. Expression of a gene family in the dimorphic fungus *Mucor racemosus* which exhibits striking similarity to human ras genes. Mol. Cell. Biol. 1990; 10: 6654-6663.

[112] Cihlar RL, Sypherd PS. Complementation of the leuB6 allele of *Escherichia coli* by cloned DNA from *Mucor racemosus*. J Bacteriol 1982; 151: 521-523.

[113] Linz JE, Katayama C, Sypherd PS. Three genes for the elongation factor ef-lx in *Mucor racemosus*. Mol Cell Biol, 1986; 6: 593-600.

[114] Doiphode N, Rajamohanan PR, Pore V, Ghormade V, Deshpande MV. Chitosan production using a dimorphic zygomycetous fungus *Benjaminiella poitrasii*: role of chitin deacetylase. Asian Chitin J 2009; 5: 19-26.

[115] Chitnis M, Munro CA, Brown AJP, Gooday GW, Gow NAR, Deshpande MV. The zygomycetous fungus, *Benjaminiella poitrasii* contains a large family of differentially regulated chitin synthase genes. Fungal Genet Biol 2002; 36: 215-223.

[116] Bulawa CE, Osmond BC. Chitin sunthase I and chitin synthase II are not required *in vivo* in *Saccharomyces cerevisiae* Proc Natl Acad Sci USA 1990; 87: 7424-7428.

[117] Silverman SJ, Sburlati A, Slater ML, Cabib E, Chitin synthetase is essential for septum formation and cell division in *Saccharomyces cerevisiae*. 1988; Proc Natl Acad Sci USA 1990; 87: 7424-7428.

[118] Munro CA, Winter K, Buchan A, Henry K, Becker JM, Brown AJ, Bulawa CE, Gow NAR. Chs1 of *Candida albicans* is an essential chitin synthase reqired for synthesis of the septum and cell integrity. Mol Microbiol 2001; 39: 1414-1426.

[119] Culp DW, Dodge CL, Miao YH, Li L, Sag-Ozkal D, Borgia PT. The *chsA* gene from *Aspergillus nidulans* is necessary for maximal conidiation. Curr Microbiol 2000; 182: 349-353.

[120] Aufavre- Brown A, Mellado E, Gow NAR, Holden DW. *Aspergillus fumigatus chsEII*: a gene related to *CHS3* of *Saccharomyces cerevisiae* and important for hyphal growth and conidiophores development but not for pathogenecity. Fungal Genet Biol 1997, 21:141-152.

[121] Kafetzopoulos D, Thireos G, Vournakis JN, Bouriotis V. The primary structure of a fungal deacetylase reveals the function for two bacterial products. Proc Nat Acad Sci USA 1993; 90: 8005-8008.

[122] Christodoulidou A, Bouriotis V, Thireos G. Two sporulation-specific chitin deacetylase-encoding genes are required for the ascospore wall rigidity of *Saccharomyces cerevisiae*. J Biol Chem 1996; 271: 31420-31425.

[123] Mishra C, Semino C, Mckreath K, De La Vega H, Jones BJ, Specht CA, Robbins P. Cloning and expression of two chitin deacetylase genes of *Saccharomyces cerevisiae*. Yeast. 1997; 13: 327-336.

[124] Maw T, Tan TK, Khor E, Wong SM. Complete cDNA sequence of chitin deacetylase from *Gongronella butleri* and its phylogenetic analysis revealed clusters corresponding to taxonomic classification of fungi. J Biosci Bioeng 2002; 93: 376-381.

[125] Jeraj N, Kunic B, Lenasi H, Breskvar K. Purification and molecular characterization of chitin deacetylase from *Rhizopus nigricans*. Enz Microbial Technol 2006; 39: 1294-1299.

[126] Wessels JGH. Cell wall synthesis in apical hyphal growth. Int Rev Cytol 1986; 104: 37-39.

[127] Bartnicki-Garcia S, Hergert F, Gierz G. A novel computer model for generation of cell shape: application to fungal morphogenesis. In: Kuhn PJ, Trinci APJ, Jung MJ, Copping LG, Eds Biochemistry of Cell Walls and Membranes in Fungi, Springer-Verlag, Berlin 1990; pp 43-60.

[128] Trinci APJ, Saunders PT. Tip growth of fungal hyphae. J Gen Microbiol 1977; 113: 243-248.

[129] Bartnicki-Garcia S, Lippman E. Fungal morphogenesis: cell wall construction in *Mucor rouxii*. Science 1969, 165: 302-304.

[130] Gooday GW. A autoradiographic study of hyphal growth of some fungi. J Gen Microbiol 1971, 67: 125-133.

[131] Prosser JI, Trinci APJ. A model for hyphal growth and branching. J Gen Microbiol 1979, 111: 153-164.

[132] Bartnicki-Garcia S, Hergert F, Gierz G. Computer simulation of fungal morphogenesis and the mathematical basis for hyphal (tip) growth. Protoplasma 1989; 153: 46-57.

[133] Walther T, Reinsch H, Ostermann K, Deutsch A, Bley T. Applying dimorphic yeasts as model organisms to study mycelial growth: Part 2. Use of mathematical simulations to identify different construction principles in yeast colonies. Bioprocess Biosyst Eng 2011; 34: 21-31.

[134] Sugiyama J. Relatedness, phylogeny and evolution of the fungi. Mycoscience 1998; 39: 487-511.

[135] Berbee ML, Taylor JW. Fungal molecular evolution: gene trees and geologic time. In: McLaughlin DJ, McLaughlin EG, Lemke PA, Eds The Mycota VII. Systematics and Evolution. Part B. Springer-Verlag, New York. 2001. pp 229-245.

[136] Voigt K, Woestemeyer J. Phylogeny and origin of 82 zygomycetes from all 54 genera of the Mucorales and Mortierellales based on combined analysis of actin and translation elongation factor EF-1a genes. Gene 2001, 270:113-120.

[137] Tanabe Y, Watanabe MM, Sugiyama J. Evolutionary relationships among basal fungi (Chytridiomycota and Zygomycota): insights from molecular phylogenetics. J Gen Appl Microbiol 2005, 51: 267-276.

[138] O'Donnell K, Lutzoni FM, Ward TJ, Benny GL. Evolutionary relationships among mucoralean fungi (Zygomycota): Evidence for family polyphyly on a large scale. Mycologia 2001; 93: 286-296.

[139] Prillinger H, Oberwinkler F, Umile C, Tlachac K, Bauer R, DorflerC, Taufratzhofer E. Analysis of cell wall carbohydrates (neutral sugars) from ascomycetous and basidiomycetous yests with and without derivatization. J Gen Appl Microbiol 1993; 39:1-13.

[140] Prillinger H. Yeasts and anastomoses: their occurrence and implication for the phylogeny of Eumycota. In: Rayner ADM, Brasier CM, Moore D, Eds Evolutionary Biology of the Fungi. Cambridge University Press, Cambridge, UK. 1987; pp 355-377.

[141] Bartinicki-Garcia S. The cell wall: a crucial structure in the fungal evolution.In: Rayner ADM, Brasier CM, Moore D, Eds Evolutionary Biology of the Fungi. Cambridge University Press, Cambridge, UK. 1987; pp 389-403.

[142] Dorfler CH, Lehle L, Prillinger H. Cell wall analysis of basidiomycetous yeasts from homobasidiomycetes. Mycol 1986; 52: 347-358.

[143] Dow JM, Rubery PH. Chemical fractionation of the cell walls of mycelial and yeast-like forms of *Mucor rouxii*: a comparative study of the polysaccharide and glycoprotein components. J Gen Microbiol 1977; 99: 29-41.

[144] Ruiz-Herrera J, Gonzalez-Prieto JM, Ruiz-Medrano R. Evolution and phylogenetic relationships of chitin synthases from yeasts and fungi. FEMS Yeast Res 2002; 1: 247-256.

[145] Bowen AR, Chen-Wu JL, Momany M, Young R, Szanislo PJ, Robbins PW. Classification of fungal chitin synthases. Proc Natl Acad Sci USA 1996; 89: 519-523.

[146] Ruiz-Herrera J, Ortiz-Castellanos L. Analysis of the phylogenetic relationships and evolution of the cell walls from yeasts and fungi. FEMS Yeast Res 2010; 10: 225-243.

[147] James TY, Kauff F, Schoch CL, *et al.* Reconstructing the early evolution of fungi using a six-gene phylogeny. Nature 2006; 443: 818-822.

[148] Riquelme M, Bartnicki-garcia S. Advances in understanding hyphal morphogenesis:Ontogeny, phylogeny and cellular localization of chitin synthases. Fungal Biol Rev 2008; 22: 56-70

[149] Gordon CL, Khalaj V, Ram AF, Archer DB, Brookman JL, Trinci AP, Jeenes DJ, Doonan JH, Wells B, Punt PJ, van den Hondel CA, Robson GD. Glucoamylase: green fluorescent protein fusions to monitor protein secretion in Aspergillus niger. Microbiol. 2000; 146, 415-426.

[150] Wosten HA, Moukha SM, Sietsma JH, Wessels JG. Localization of growth and secretion of proteins in Aspergillus niger. J. Gen. Microbiol. 1991; 137, 2017-2023.

[151] Wolff AM, Arnau J. Cloning of Glyceraldehyde-3-phosphate Dehydrogenase-Encoding Genes in *Mucor* circinelloides (syn. racemosus) and use of the gpd1 promoter for recombinant protein production. Fungal Genet. Biol. 2002; 35, 21-29.

[152] Bredenkamp A, Velankar H, Van Zyl WH,Gorgens JF. Effect of dimorphic regulation on heterologous glucose oxidase production by *Mucor* circinelloides. Yeast 2010; 27: 849-860.

[153] Phadtare S., Rao M. and Deshpande V. (1997) A serine alkaline protease from the fungus Conidiobolus coronatus with a distinctly different structure than the serine protease subtillisin Carlberg. Arch. Microbiol.166, 414-417.

[154] Ghormade V, Kulkarni S, Doiphode N,. Rajmohanan PR , Deshpande MV. Chitin deacetylase: A comprehensive account on its role in nature and its biotechnological applications.In: Méndez-Vilas A, Ed. Current Research, Technology and Education Topics in Applied Microbiology and Microbial Biotechnology. Spain. 2010; pp 1054

[155] Vicente G, Bautista LF, Rodrı́guez R, Gutierrez FJ, Sadaba I, Ruiz-Vazquez R M, Torres-Martınez S, Garre V.. Biodiesel production from biomass of an oleaginous fungus. Biochem Eng J 2009; 48: 22-27.

[156] Binupriya AR, Sathishkumar M, Yun SI. Biocrystallization of silver and gold ions by inactive cell filtrate of *Rhizopus* stolonifer. Colloids Surfaces B: Biointerfaces 2010;79: 531-534

[157] Nakayashiki H, Nguyen QB.RNA interference: roles in fungal biology. Current Opinion in Microbiology 2008, 11:494-502

[158] Nicolas FE, de Haro JP, Torres-Martinez S, Ruiz-Vazquez RM. Mutants defective in a *Mucor* circinelloides dicer-like gene are not compromised in siRNA silencing but display developmental defects. Fungal Genet Biol 2006, 44:504-516.

CHAPTER 8

Development and Dimorphism of the Phytopathogenic Basidiomycota *Ustilago maydis*

José Ruiz-Herrera[*] and Claudia G. León-Ramírez

Departamento de Ingeniería Genética, Unidad Irapuato. Centro de Investigación y de Estudios Avanzados del IPN, México. Km. 9.6 Libramiento Norte Carretera Irapuato-León. CP. 36500 Apartado Postal 629, Mexico

Abstract: *Ustilago maydis*, a Basidiomycota species, is the causal agent of common smut in corn and teozintle. This fungus has a complex life cycle regulated by two mating factors: *a* with two alleles involved in mating, and *b*, multiallelic, involved in mycelial growth and pathogenesis. Two morphological stages can be recognized in the life cycle of the fungus: a yeast-like haploid, saprophytic stage (sporidia), and a mycelial dikaryotic pathogenic stage. Transition of the first one into the latter involves mating between two sporidia containing different *a* and *b* alleles, only these dikaryotic cells being infectious, and maintained only in the host, where the diploid stage, a specialized type of spore (teliospore) is formed after cytokinesis. Teliospores germinate to produce haploid sporidia. It is known that pathogenesis and morphogenesis are controlled by the formation of a heterodimer made of two gene products from compatible *b* genes (bx/by), that acts as a master transcription factor. Nevertheless new data have demonstrated that *U. maydis* haploid cells may be pathogenic to different plant species under experimental conditions, and that they may grow in the lab in the mycelial form responding to different stimuli: nitrogen starvation, use of fatty acids as C source, and incubation at acid pH. All these data suggest that both, pathogenesis and morphogenesis may occur without the involvement of the heterodimer through a bypass involving a MAPK pathway which is antagonized by a PKA pathway. Different elements such as polyamines, probably DNA methylation, cell wall alterations, but not the pH responsive Pal/Rim pathway are involved in these processes.

Keywords: *Ustilago maydis*, dimorphism, morphogenesis, mating type, pathogenicity, cAMP, hyphae. yeast-like, cell wall, differentiation, virulence, tumor development, teliospores.

INTRODUCTORY ASPECTS

Ustilago maydis (DC.) Cda., a basidiomycota fungus, is the causal agent of corn smut or "huitlacoche", a disease that may cause important economical losses under some conditions, but paradoxically, infected sweet corn constitutes a delicacy in central México, with the result that an infected cob increases its commercial value.

In recent years, *U. maydis* has become an attractive model for the study of basic problems of fungal biology, such as differentiation, the genetic control of mating, dimorphism and the interactions of host and pathogen in plant fungal diseases [1, 2]. In this last aspect *U. maydis* constitutes a reliable and safe model for comparative studies with other members of pathogenic *Ustilaginales* that constitute true pests in the field, having the potential to cause severe disease outbreaks despite chemical treatment, the existence of partially resistant cultivars, and the different plough methods employed. Among them we can cite: *Ustilago hordei, Ustilago nuda, Ustilago nigra, Tilletia indica, Tilletia caries* and *Tilletia controversa*.

U. maydis provides an ideal experimental model because of its facility to grow in the laboratory, the existence of a sexual cell cycle that permits genetic analysis, and the availability of genetic and molecular tools for its study. There exist numerous selective markers for *U. maydis*, reporter genes, and constitutive promoters [3-6]; it can be easily transformed and its high level of homologous recombination allows to obtain mutants by gene disruption [7, 8]. The existence of diploids facilitates the analysis of dominance.

*****Address correspondence to José Ruiz-Herrera:** Departamento de Ingeniería Genética, Unidad Irapuato. Centro de Investigación y de Estudios Avanzados del IPN, México. Km. 9.6 Libramiento Norte Carretera Irapuato-León. CP. 36500 Apartado Postal. 629; Tél: +5246239600; Fax: +524626235948; E-mail: jruiz@ira.cinvestav.mx

During its meiotic process it is possible to perform segregation analyses. Besides, its life cycle can be completed in maize germlings in a matter of about three weeks [9], and finally, its genome has been sequenced and annotated [10]; MUMDB data base; http://www.mips.gsf.de/genre/proj/ustilago/.

In nature *U. maydis* has a restricted host range infecting only maize (*Zea mays* L.), and its putative ancestor teozintle (Fig. **1**) (*Zea mays* spp. *parviglumis*). *U. maydis* is not an obligate parasite, but requires its natural plant host to complete its sexual cycle. Owing to all these characteristics, the study of the fungus has given rise to extensive reviews on its different characteristics: sexuality, infection of maize plants, its pathogenic behavior, *etc.* [11-17]. In the present chapter we will briefly discuss only these aspects of the fungus, and will dedicate a deeper discussion to the physiology and biochemical mechanisms related to dimorphism, according to the general theme of the volume.

Figure 1: Infection of teozintle (*Zea mays* spp *parviglumis*) by *Ustilago maydis*. A rare photograph of teozintle infection in the field. Photograph by Cristina G. Reynaga and J. Ruiz-Herrera.

Life Cycle of *U. maydis*

The life cycle of *U. maydis* is very complex with alternating haploid and diploid stages (Fig. **2**). Although the fungus is not a strict parasite, it strictly requires the host to complete its sexual cycle. In its saprophytic stage, *U. maydis* grows in the form of cigar-shaped yeast cells, measuring 12-18 μm in length and 4-5 μm in width [18, 19]. These cells reproduce by budding, characteristically appearing at one end of the cell forming a typical angle with the mother cell. When two yeast cells belonging to complementary groups are close enough, they start forming slender tubular structures called mating tubes that direct to each other until they fuse, typically at the tip. At this time, a tubular structure normally thicker than the conjugation tubes appears at the junction point, and the nuclei from the original cells migrate to this hypha, named germination tube or dikaryotic hyphae. This dikaryotic stage is unstable outside the host, except under some special conditions, mainly under nitrogen starvation [20] that grows apically until it can invade a susceptible host. The hyphae penetrate into the plant, normally forming an appresorium, although penetration through wounds is possible. The infectious hyphae grows on the surface of the cells before penetrate them [21], and when it does it may grow into them between the wall and the cytoplasm without causing damage to the cytoplasm. The growing hyphae in the plant tissues suffers a series of morphological alterations, including branching, formation of a mucilaginous material, development of osmotic sensitive cells before giving rise to the teliospores, where karyogamy takes place, constituting the diploid phase of the cell cycle [20]. Teliospores are quasi spherical, have a different cell wall thicker than the one of the yeast or mycelial forms, are melanized and echinulate measuring 7-8 μm in diameter. It is important to recall that teliospores are formed only in its natural hosts, probable because their formation requires specific signals from the plant. When the teliospores are liberated to the medium, they germinate under adequate conditions with the formation of a short filament named promycelium of a length of 20-25 μm and 2-3 μm wide [22]. Meiosis occurs with the formation of four basidiospores, that fall off from the promycelium and start a new life cycle.

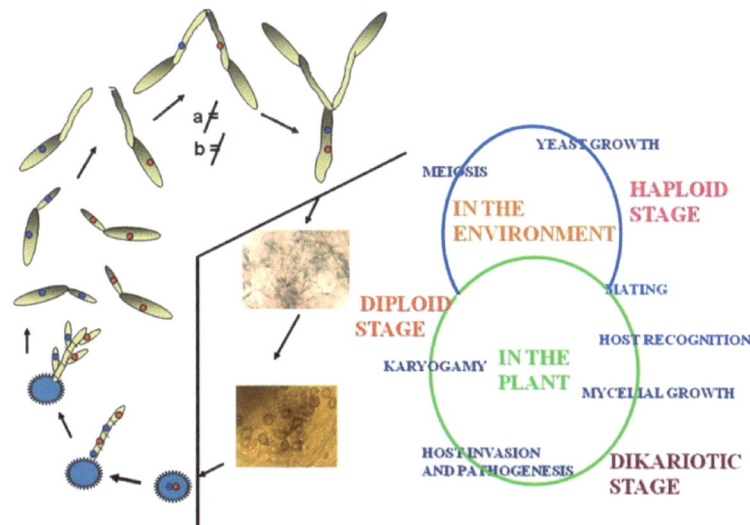

Figure 2: Life cycle of *Ustilago maydis*. Left, schematic stages of the life cycle. Germination of teliospores with formation of the promycelium (basidium) with formation of four basidiospores. These detach from the promycelium, germinate, and reproduce by budding, until two sexually compatible cells mate and develop a dikaryon. This grows in the mycelial form, invades a suitable host, and finally after caryogamy takes place, it forms diploid teliospores to close the life cycle. Right, schematic representation of the two stages of the life cycle (saprophytic and virulent), and the cellular forms of the fungus.

Mating in *U. maydis*

When two sexually-compatible strains of *U. maydis* are mixed and inoculated on solid medium they grow in the form of white cotton-like colonies, contrasting with the smooth colonies formed by haploid strains or incompatible mixtures. This mycelial growth has been named FUZ phenotype [23, 24]. This phenotype has been used as a test to identify the mating type of isolates, using tester strains of known sex.

The process of mating in *U. maydis* is tetrapolar with the intervention of two mating *loci* denominated *a* and *b*. There exist two alleles, *a1* and *a2* (better considered to be idiomorphs because of their sequence differences, but named alleles indistinctly) containing two ORFs of a size of 4.0 and 8.5 Kb respectively [22]. One of these ORFs encodes a pheromone specific for each allele (*Mfa1, Mfa2*), and the other a specific receptor for the opposite pheromone (*Pra1, Pra2*). The peptides *mfa1* and *mfa2* are 38 and 40 amino acids in length respectively and are processed with the formation of the pheromones made of 9 and 12 amino acids respectively that contain a *C*-terminal sequence *Caa*, characteristic of pheromones of other fungi, that becomes prenylated for its function. *Pra1* and *Pra2* are made of 357 and 346 amino acids respectively, and contain 7 transmembrane domains, and as mentioned above they act as receptors of the opposite pheromones, *Pra1* of *Mfa2* and *vice versa*.

There are more than 30 *b* ideomorphs identified in nature, each one with two ORFs that encode two different polypeptides whose reading is divergent and are named *bE* and *bW* [18, 25, 26]. The encoded polypeptides are made of 426 and 626 amino acids respectively and have no significant homology among them, but all have a *C*-terminal region with high homology between the different bW or bE polypeptides, a central homeodomain, and a *N*-terminal highly variable. The role of *b* genes depends on their formation of a heterodimer that acts as a transcription factor. Formation of the heterodimer occurs only between *bW* and *bE* polypeptides when these come from different alleles: *bEx* X *bWy*.

Mating between two yeast cells or sporidia occurs only if they carry different *a* and *b* alleles (*a1 bx* X *a2 by*; [27, 28]). It is known that *a* genes are involved in mating and maintenance of the mycelial growth; whereas *b* genes are involved in mycelial growth, infection of the plant and development of the symptoms. Mating occurs by the recognition of the corresponding pheromone by the receptors from sporidia of the opposite mating type (see above). This process gives rise to G2 arrest of the mating partners, and to the transfer of the information through two signal transduction pathways, one PKA and one MAPK, that simultaneously converge in the transcription factor Pfr1 (for a review see [29]), starting an autocrine activation that gives rise to a stimulation of transcription, besides of *MFA* and *PRA*, of other genes including *b* genes.

Pathogenicity of *U. maydis*

As mentioned before, only the heterokaryon formed as a result of mating between sexually compatible sporidia is infectious in maize or teozintle. It may be thus concluded that the sexual cycle, virulence and morphogenesis are strictly correlated phenomena in *U. maydis*. Nevertheless, under experimental conditions, the association between these phenomena can be altered facilitating their study. Accordingly, forced diploids produced by the fusion of compatible sporidia are also infectious, although they are somewhat less virulent than dikaryons; even more, artificially obtained merodiploids carrying both *a1/a2* and *bx/by* genes are also virulent. And also as above indicated, against the concept that only maize and teozintle are the natural susceptible hosts for infection, under axenic experimental conditions *U. maydis* can infect plants belonging to different taxonomic groups [30, 31].

Infection of the host occurs as indicated above by penetration of the dikaryotic mycelium through natural or artificial openings, but mostly by formation of appresoria, a process that apparently is stimulated by a hydrophobic surface, depends on the transcriptional regulator Biz1 [32]; and involves the specific MAPK Kpp6 [33]. Most probably, penetration of the plant epidermis involves hydrolytic enzymes [34]. Once in the interior of the host, *U. maydis* establishes a compatibility with the plant leading to a biotrophic type of infection; the limited battery of hydrolytic enzymes being used only for penetration, but not for tissue destruction [35]. The pathogen grows almost exclusively in the plant meristems, *i.e.,* in actively growing tissues. The most common characteristics of the infection are increase in the formation of anthocyanins, appearance of chlorosis, and most important, the formation of galls or tumors, where teliospores accumulate, giving the characteristic dark color to these, from where the name smut is derived. In heavily infected plants, dwarfing and even death may occur.

The pathogenic behavior of the organism depends on the bE/bW heterodimer (see above) that in response to selective phosphorylation by Pka and the MAPK Kpp2 regulates 347 genes, 212 upregulated and 135 downregulated involved in a number of physiological activities including lipid metabolism, cell cycle control, and DNA replication [17]. It appears that the control of all these genes by the heterodimer is not direct, but indirect, probably through the activation or inactivation of secondary transcription factors. One exceptional feature of *U. maydis* (the first case in fungal pathogens) is that groups of genes involved or related to infection appear in the form of 12 clusters [10], whose products are upregulated or downregulated during infection. Infection with mutants with one of these clusters deleted display reduced virulence, in one case with even defects in teliospore formation, but surprisingly, in one case virulence is increased [10].

Interestingly, it appears that *U. maydis* represses apoptosis in the host cells, a feature known to be involved in the formation of reactive oxygen species (ROS), obviously an important negative feature for a biotrophic pathogen. As a response to ROS production the fungus reacts with mechanisms of ROS detoxification, apparently controlled by the activity of a gene denominated *YAP1* mainly through peroxidase and S-glutathione transferases [36, 17]. It is known that biotrophic pathogens induce compatible interactions through jasmonic acid (JA), while necrotrophic ones induce the salicylic acid-dependent defense mechanisms in plants. As would be expected, *U. maydis* infection of maize induces activation of JA-dependent products, and none of those induced in necrotrophic responses [35, 17].

In advanced stages of the infection, tumors or galls are produced accompanied by induction of auxins and genes regulated by auxins [35]. This process appears to be dependent of the host, and not of the pathogen

[37], since *U. maydis* mutants deleted in important genes involved in the synthesis of indolacetic acid are able to induce tumor formation. Nevertheless this process occurs most probably in response to *U. maydis* inducers. The response of the host to *U. maydis* infection is not due to a gene-for-gene intraction, but to a mixture of different factors. It has been demonstrated, that among these, a mechanism of protection by the plant is an increased production of polyamines [38]; it is known that polyamines constitute a general defense mechanism of plants against fungal infections.

A very important discovery to explain the mechanism of *U. maydis* pathogenicity, was the observation by use of a transcriptome profiling analysis, that the number and identity of the different pathogen and maize genes whose expression was regulated during invasion of the plant, were strictly tissue-dependent [39]. These results suggest that the fungus utilizes different mechanism of infection depending on the invaded tissues of the host.

As was insistently indicated, the natural hosts of *U. maydis* are maize and its putative ancestor, teozintle. Nevertheless, we found that under axenic conditions *U. maydis* had the ability to infect a number of species, completely unrelated to maize, both monocts and dicots: gingko, sorghum, rice, garlic, tobacco, potato, bean, African violet and papaya. The infected plants developed distinct symptoms similar to those seen in infected maize plants: development of mycelium, chlorosis, elevated anthocyanins synthesis, necrosis, and stunting. The developmental changes in some species induced by *U. maydis* were development of adventitious roots and specially tumor formation in papaya plants [30]. Interestingly, these symptoms could be induced by infection with haploid strains, but as expected inoculation with a diploid strain did not give rise to the formation of teliospores; *i.e.*, the sexual cycle can be completed in the natural hosts only. Further experiments carried out on the infection of *Arabidopsis thaliana* demonstrated that even a number as low as 10 cells of *U. maydis* inoculated per plant, gave rise to the symptoms described above. As occurred with other plant species, the haploid strains were virulent, and when plants were inoculated with diploids, no teliospores were formed [31]. Microscopic examination of infected plants demonstrated the presence of mycelium growing in the tissues of the host, and some images showed penetration of the fungus through stomata, with formation of putative appresoria. According to these results it appears that as occurs with the dimorphic transition (see below) there exist alternative routes to the normal pathogenic behavior that are independent of the heterodimer (see above). Also interesting was the observation that mating and genetic recombination might occur without the direct establishment of an infectious process. When *U. maydis* was grown on top of maize calli separated by a membrane, genetic recombination occurred, and structures similar to teliospores were formed, although these were unable to germinate [40].

The Phenomenon of Dimorphism in *U. maydis*

General Aspects

As described above, *U. maydis* in its saprophytic stage grows in a yeast-like budding form, and changes to a mycelial growth, that is the invasive stage, when two sexually-compatible yeasts (sporidia) mate to form a dikaryon. Under other both natural or artificial conditions, the fungus can grow in a mycelial form, for example a prolonged hyphal growth is obtained when the dikaryon grows in low nitrogen conditions; forced diploids obtained by fusion of sexually compatible strains carrying different auxotrophic markers and selection on minimal media may be stable and display a FUZ phenotype, merodiploids containing different *b* idiomorphs grow in a mycelial phase; different haploid mutants may grow constitutively in the mycelial form; such as adenylate cyclase minus mutants (*uac1*) [41] and a mutant defective in a histone deacetylase (J. M. González-Prieto and J. Ruiz-Herrera, unpublished), incubation of the fungus in the presence of lipids or fatty acids [42], and incubation in media with a low pH [43]. These last data demonstrate that the formation of an *ax/by* heterokaryon is not necessary for mycelial growth, and that other pathways exist for the dimorphic phenomenon. As a further evidence of this assert, it was demonstrated that *b* minus mutants behaved the same as the wild type strains when grown at acid pH [43].

Role of Signal Transduction Pathways

In *U. maydis* two signal transduction pathways, one an AMPc dependent PKA pathway and a MAPK pathway have been demonstrated to be involved in mating and virulence (reviewed in [44-46]). Under natural conditions these phenomena are under the control of the transcription factor *Prf* that is differentially phosphorylated by the PKA *Adr1*, and the MAPK *Kpp2* [17]. Mutation of the genes encoding the adenylate cyclase (*UAC1*) or the catalytic subunit of the PKA (*ADR1*) give rise to a mycelial constitutive phenotype, whereas mutants in the gene encoding the regulatory PKA subunit (*UBC1*) grow constitutively in the yeast form [41, 47]. These results were confirmed by the observation that addition of cAMP to cultures of wild type or morphological mutants, with the exception of *adr1* mutants, repressed mycelial growth, independently of the pH of the medium [48]. Interestingly, mutants defective in the gene that encodes the minor PKA catalytic subunit (*UKA1*) grow constitutively in the yeast form, and double *adr1uka1* mutants grow in the mycelial form. These results were interpreted as evidence that Ard1 inhibits mycelial growth, whereas Uka1 antagonizes this effect and stimulates filamentation [47]. In conclusion, it appears that whereas both pathways act synergically in mating and virulence, they operate antagonically on dimorphism, the PKA pathway being involved in yeast growth, and the MAPK pathway in mycelial development.

Nitrogen Starvation and Dimorphism

Regarding the hyphal growth of *U. maydis* in response to reduced ammonium or poor nitrogen sources in the growth medium, the intervention of the PKA signal transduction pathway, as well as the existence of a role for the ammonium transporters Ump1 and Ump2 have been considered. The observation that Ump2, complemented the mycelial growth of *U. maydis ump1, ump2, mp3* mutants suggested that this was the sensor of ammonium concentration in the medium, and the connection with the PKA signal transduction pathway during hyphal development [15].

Dimorphic Transition Control by pH

In a similar way to other fungi like *Candida albicans*, the pH of the medium has influence on the morphological development of *U. maydis*, except that the response of these two fungi is opposite; whereas mycelium growth in *C. albicans* occurs at neutral, but not at acid pH, mycelial development of *U. maydis* requires acid pH [43]. In this case the nitrogen source proved to be important, since besides an acid pH, the dimorphic transition required the use of NH_4Cl as N source, since in the use of NH_4NO_3 repressed mycelial growth, probably because in this case pH increased during incubation, while in the presence of NH_4^+, the operation of a H^+/NH_4^+ antiport tended to maintain an acid pH. Important also was to take the cells to G_0 by a period of starvation. The dimorphic transition from mycelium to yeast was found to be freely reversible, but the opposite transition was reversible only during the initial stage of the growth phase [43]. As described above, mycelial growth induced by acid pH was displayed not only by diploid strains, but also by haploid strains, and by mutants in *b* genes, *bW* or *bebW*, revealing that the heterodimer and its signal transduction pathway were not involved in the process [43]. As indicated above, addition of cAMP repressed mycelium formation by the wild strain at pH 3, revealing an inhibitory function of the PKA signal transduction pathway on the dimorphic transition activated by acid pH. This hypothesis was confirmed when it was observed that an *ubc1* mutant (defective in the regulatory subunit of the PKA) grew in the yeast form at acid pH while on the other hand, mutants in members of the MAPK pathway grew in the yeast form, confirming the antagonic roles of PKA and MAPK pathways in the dimorphic transition induced by acid pH [48].

Lipids and Dimorphism

A surprising result in the study of *U. maydis* dimorphism was the observation that wild type haploid or diploid strains were able to grow in the form of mycelium when incubated with any plant oil as carbon source. This effect was shown to be due to triglycerides, since the addition of triolein produced the same effects. Moreover fatty acids and even tweens of different chain length and saturation were equally effective as morphogenetic inducers, but whereas diploids appeared fully mycelial, haploid cells formed shorter filaments like pseudomycelium. Glycerol also induced mycelial growth, but hyphae were shorter, whereas glucose, even in the presence of fatty acids reverted the hyphal growth to yeast-like [42]. The observation that arabinose and only trace amounts of tween40 induced hyphae formation, suggested that fatty acids acted as morphogenetic signals. On solid media with tween40, invasive branching mycelial growth was obtained. The observation that *ubc* mutants that grow constitutively in the yeast-like form, and mutants in members of the MAPK pathway do not respond to fatty acids suggest an antagonic role of PKA and MAPK signal transduction pathway [42], exactly as occurs with the normal and pH dependent morphogenesis (see above). In an attempt to determine whether lipid metabolism was required for lipid-induced dimorphic transition and pathogenesis, mutants lacking the *mfe2* gene that encodes the multifunctional enzyme for the second and third steps in peroxisomal β-oxidation of fatty acids were isolated [49]. It is known that peroxisomes are involved in the metabolism of long and very long fatty acids, accordingly the authors determined the *in vitro* growth characteristics in the presence of fatty acids of different chain length. They observed that while C_{20} or C_{12} fatty acids induced mycelial growth of the wild type strain, the fatty acids of very long chain length were unable to induce mycelial growth of the mutant, indicating the necessity of fatty acid oxidation for their morphogenetic effect. Maize inoculation with the mutants showed decreased pathogenicity; data that reveal the role of the metabolism of these fatty acids in virulence [49].

Polyamines and Dimorphism

Polyamines are known to be essential for growth of all living organisms, and also required in a quantitative dependent manner for development and differentiation. Previously we described that polyamines are involved in several differentiation phenomena, by the observation that preceding every differentiation step occurred an elevation in their concentrations, and in ornithine decarboxylase ODC [50, 58] synthesis. All these phenomena were blocked by diaminobutanone (DAB), an inhibitor of ODC [50, 51]. In line with these results we observed that DAB inhibited teliospore germination and dimorphic transition in *U. maydis* [52]. Cloning and mutation of the gene encoding ODC from *U. maydis* permitted a more direct test of the requirement of polyamines in its dimorphic transition. These mutants were auxotrophic for polyamines, but if supplemented with 0.5 mM putrescine they grew at the same rate as the wild type, but in a yeast-like form at acid pH. Only when supplemented with 4-5 mM putrescine, they grew in the mycelial form at pH 3 [52]. These results are strong evidence that polyamines are important for cell differentiation in *U. maydis*. More recently, isolation and mutation of the gene encoding spermidine synthase (Spe), demonstrated the same concentration-dependent requirement of spermidine to sustain mycelial growth at acid pH. Spermidine resulted the most important polyamine for *U. maydis* growth and differentation [53]. Interestingly, Spe is encoded by a chimeric gene that also has an ORF for saccharopine dehydrogenase, an enzyme involved in lysine biosynthesis. In this respect, we found that the existence of a homologous chimeric gene is specific of Basidiomycota [54]. As occurs with all systems, the exact mechanism of polyamine function in cell differentiation remains unknown, although it has been hypothesized that it plays a role in the selective activation of some genes involved in these processes.

Differential DNA Methvlation

DNA methylation is an epigenetic mechanism for the control of gene expression. In eukaryotic cells 5-methylcytososine is the only methylated base found in significant amounts in DNA. Postsynthetic methylation of DNA is catalyzed by different methylases, that using *S*-adenosylmethionine as substrate introduce a methyl residue into selected cytosine residues of DNA. High levels of DNA methylation are associated to low transcriptional activity and silent chromatin, whereas low methylation levels are associated with high transcriptional activity. Available data revealed that DNA methylation is involved in different physiological activities, such as carcinogenesis differential gene expression, timing of DNA replication, inactivation of chromosme X and genomic imprinting ([55] and references cited therein), and the number of fungal species where DNA methylation has been demonstrated is scant [51, 56, 57], and these include *U. maydis* [58], although *in silico* studies have not revealed the existence of a DNA methtyltraferase, probably because of the extremely low conservation of their sequences. Previous data suggested a role of DNA methylation in fungal cell differentiation; accordingly, when we adapted the technique of Amplified Fragment Length Polymorphism (AFLP) to analyze differential DNA methylation, we were able to demonstrate the existence of significant differences between the yeast and mycelial forms of dimorphic fungi representative of the main fungal taxa: *Mucor rouxii; Yarrowia lipolytica* and *U. maydis* [55] suggesting a relationship between DNA methylation and fungal differentiation. Further work demonstrated that in these fungi, some of these differences were strictly correlated to the dimorphic transition [59].

As described above, a possible association between polyamine metabolism was observed (see above). Looking for the mechanism of this effect, it was demonstrated that 5 azacytine, a cytosine analogue that interferes with DNA methylation relieved the inhibition of DAB (see above) on the germination of sporangiospores of *M. rouxii* [60]. Moreover, it was observed that transcription of specific genes during spore germination occurred only after DNA became demethylated, and the DAB inhibited their transcription [61]. In further experiments it was demonstrated that polyamines, especially spermidine, inhibited different cytosine DNA methylases, whereas they had no effect on adenosine DNA methylases or restriction enzymes that recognized the same sequences than the former [62].

The Pal/Rim Pathway

A signal transduction pathway that is involved in the response of fungal cells to changes in pH was originally described in *Aspergillus nidulans* and *S. cerevisiae*. The mechanism of operation of this pathway leads to the proteolytic activation of a transcription factor named PacC in *A. nidulans* and Rim101 in *S. cerevisiae*, that contains three zinc finger-DNA binding domains and a regulatory domain that is eliminated by proteolysis. Transfer of the external pH signal depends on 5 or 6 proteins denominated PalH/Rim21, PalF/Rim8, PalI/Rim9, PalA/Rim20, PalB/Rim13 and PalC/Rim23 (for the sake of facility we have denominated the pathway as Pal/Rim), that form two complexes, one located at the plasma membrane and constituted by PalA, PalB and transitorily by PalC, that is responsable for extracellular pH sensing; and the other one located at the endosome membrane, constituted by PalA and PalB, involved in the proteolytic activation of PacC [63-65]. This pathway was considered to be specific of *Ascomycota* species, until we described the presence of a homologue of PacC *in U. maydis*, and moved our interest to know whether this pathway was involved in the dimorphic regulation of the dimorphic transition induced by acid pH in this fungus [66]. Further work led to the demonstration that the fungus contained homologues of the members of the endocytic membrane, but not of the members of the complex located at the plasma membrane [67]. Interestingly, this appears to be a general characteristic of the *Basidiomycota* species, some of which also lack a homologue of PalC. The nature of the receptor of the external pH signal in *Basidiomycota* thus is unknown. Taking into consideration that the Pal/Rim signal transduction pathway has been found to regulate dimorphism in response to external pH and pathogenicity in the human pathogen *Candida albicans* [68, 69], we considered that it might be also involved in regulating the dimorphic phenomenon *in U. maydis*. Disappointingly we found that mutants in every one of the homologues of the pathway in this fungus were not affected in the dimorphic transition, although they had complex phenotypic alterations including morphological alterations, sensitivity to high pH and stress, alterations in the cell wall composition and structure and secretion of a particular polysaccharide [66, 67]. Accordingly, it appears that in *U. maydis* transfer of the pH signal depends on the PKA and MAPK pathways (see above), but the nature of the pH signal receptor remains unknown.

The Cell Wall

It is known that at the end, the form of fungal cells is provided by the cell wall. Accordingly we have analyzed the structure of the cell wall of *U. maydis*, and the regulation of some biosynthetic enzymes in its yeast-like and mycelial forms. It was observed that yeast and mycelial walls had a similar general composition, with neutral polysaccharide constituting the major part of the dry weight, but slightly higher in the yeast than in the mycelial walls, 73.1 and 62.2 % respectively, whereas chitin was slightly higher in the mycelial walls, 14.2 and 16.4% respectively. Proteins accounted to 15.9% in the yeast wall and 11.7% in the mycelial wall [70]. No chitosan was detected in the wall of the yeast or mycelial cells, an odd observation, since it has been described that the *U. maydis* genome contains a substantial number of genes encoding chitin deacetylases, known to be involved in chitosan biosynthesis [16]. It is possible that chitosan is present in the cell wall at other stages of the fungus. Significant differences in the qualitative and quantitative composition of neutral sugars were also observed. Cell walls from both morphologies contained glucose, galactose, mannose ands xylose, although in different proportions, but yeast wall contained in addition ca. 6% of fucose, arabinose and rhamnose. Most proteins from either stage were solubilized by hot SDS extraction, followed by release of reduced amounts by glucanase and chitinase treatment, leaving a substantial insoluble residue. Antibodies raised against the SDS-extracted residue from the walls of yeast or mycelial walls recognized some common and specific proteins, the most noticeable difference being a 40 kDa protein that was present in the mycelial walls only [70].

It has been shown that *U. maydis* contains 8 genes encoding chitin synthases, [16], all of which have been cloned and mutated [71-75]. The most interesting phenotype corresponded to *chs6* mutants that showed strong alterations in morphology and chitin distribution, and were avirulent, although they were able to grow in the form of mycelium, but hypha were shorter and distorted [74]. Also *chs7* and *chs8* (also called *mcs1*) mutants had morphological alterations, and *chs8* was avirulent, being affected in penetration to the host tissues. Interestingly *U. maydis* hyphae were shown to display a cap of chitin, and Chs8 was found located at the hyphal apex, but *chs8* mutants lacked not only the apically located enzyme, but also the chitin cap [75]. These data suggest that Chs6 and Chs8 (both belonging to Class V) are the most important ones for *U. maydis* pathogenesis and apical growth, but their single mutants are not totally affected in dimorphism, probably because either, they can substitute each other, or because other Chs may play a minor role in mycelial growth.

It is accepted in general that the difference in growth pattern between yeast and hyphal cells is the absence or presence respectively of polar cell wall growth. This was originally evidenced a long time ago by Bartnicki-Garcia and Lippman [76], who demonstrated the differential deposition of nascent chitin in *M. rouxii* spores germinating in the yeast-like or mycelial forms. Modern studies have demonstrated that polarization is due to a number of cell components, the role of which has been extensively studied. Among these we may cite, the vesicles and microvesicles that transport enzymes and wall products to the sites where cell wall is synthesized, the cytoskeleton and its associated enzymes involved in the movement of those organelles, different GTPases, both heterotrimeric and monomeric, and the Spitzenkörper, a peculiar organelle made of microtubules, actin, and the vesicles and microvesicles located in the apical portion of growing hyphae [77-79]. In this sense, most interesting are the *in silico* analysis from Banuet *et al.* [80] that demonstrated the existence of homologues of all the proteins suggested to be involved in growth polarization in *U. maydis*. These data are evidence that transition of the yeast-like form to mycelium in *U. maydis* must involve the activation of the polarization mechanisms that lead to hyphal growth, whereas the transition from mycelium of yeast would the deactivation of this mechanism to stop or reduce polar growth.

ACKNOWLEDGEMENTS

The original work of the authors here reported was partially supported by Consejo Nacional de Ciencia y Tecnología, México. JRH is Emeritus investigator of the Sistema Nacional de Investigadores, México.

REFERENCES

[1] Bölker M. *Ustilago maydis*-a valuable model system for the study of fungal dimorphism and virulence. Microbiology 2001; 147:1395-1401.

[2] Basse CW and Steinberg G. *Ustilago maydis,* model system for analysis of the molecular basis of fungal pathogenecity. Mol Plant Pathol 2004; 5: 83-92.

[3] Botín, Kämper J and Kahmann R. Isolation of a carbon source - regulated gene from *Ustilago maydis*. Mol Gen Genet 1996; 25: 342-352.

[4] Spellig T, Bottin A, Kahmann R. Green fluorescent protein (GFP) as a new vital marker in the phytopathogenic fungus *Ustilago maydis*. Mol Gen Genet 1996; 252: 503-509.

[5] Brachmann A, Weinzierl G, Kämper J and Kahmann R. Identification of genes in the *bW/bE* regulatory cascade in *Ustilago maydis*. Mol Microbiol 2001; 42:1047-1063.

[6] Brachmann A, Konig J, Julius C and Feldbrügge M. A reverse genetic approach for generating gene replacement mutants in *Ustilago maydis*. Mol Genet Genomics 2004; 272: 216-226.

[7] Tsukuda T, Carleton S, Fotheringhan S and Holloman WK. Cloning and disruption of *Ustilago maydis* genes. Mol Cell Biol 1988; 8: 3703-3709.

[8] Fotheringham S, Holloman WK. Pathways of transformation in *Ustilago maydis* determined by DNA conformation. Genetics 1990; 124: 833-843.

[9] Banuett F. Genetics of *Ustilago maydis,* a fungal pathogen that induces tumors in maize. Genetics 1995; 29: 179-208.

[10] Kämper J, Kahmann R, Bölker M, Ma LJ, *et al.* Insights from the genome of the biotrophic fungal plant pathogen *Ustilago maydis*. Nature 2006; 444: 97-101.

[11]   Banuett F and Herskowitz I. *Ustilago maydis* smut of maize. In: Genetics of Phytopathogenic Fungi G.S. Sidhu ed. Advances in Plant Pathology 6 Academic Press London 1998; pp. 427- 455.

[12]   Ruiz-Herrera J, Martínez-Espinoza A. The fungus *Ustilago maydis,* from the aztec cuisine to the research laboratory. Int Microbiol 1998; 1: 149-158.

[13]   Ruiz-Herrera J, León-Ramírez C and Martinez-Espinoza AD. Morphogenesis and pathogenesis in *Ustilago maydis.* Recent Res Develop Microbiol 2000; 4: 585-596.

[14]   Feldbrüge M, Kämper J, Steinberg G and Kahmann R. Regulation of mating and pathogenic development in *Ustilago maydis.* Curr Opin Microbiology 2004; 7: 666-672.

[15]   Klosterman SJ, Perlin MH, Garcia-Pedrajas M, Covert SF, and Gold SE. Genetics of morphogenesis and pathogenic development of *Ustilago maydis.* Adv Genet 2007; 57: 1-47.

[16]   Ruiz-Herrera J, Ortiz-Castellanos L, Martínez AI, León-Ramírez C and Sentandreu R. Analysis of the proteins involved in the structure and synthesis of the cell wall of *Ustilago maydis.* Fungal Genet Biol 2008; 45: 571-576.

[17]   Brefort T, Doehlemann G, Mendoza-Mendoza A, Reissmann S, Djamei A, and Kahmann R. *Ustilago maydis* as a Pathogen. Annu Rev Phytopathology 2009; 47: 423-445.

[18]   Holliday R. In Handbook of Genetics, King, R. C. Ed Plenum, New York 1974; Vol 1, pp. 575-595.

[19]   Christensen JJ. Corn smut caused by *Ustilago maydis.* Monograph 2. St. Paul, MN, USA: American Phytopathological Society 1963; 41p.

[20]   Banuett F and Herskowitz I. Discrete developmental stages during teliospore formation in the corn smut fungus, *Ustilago maydis.* Development 1996; 122: 2965-2976.

[21]   Snetselaar KM. Microscopic observation of *Ustilago maydis* mating interactions. Exp Mycol 1993; 17: 345-355.

[22]   Ramberg JE and McLaughlin D. Ultrastructure study of promycelial development and basidiospore initiation in *Ustilago maydis.* Can J Bot 1980; 58: 1548-1561.

[23]   Day PR and Anagnostakis LJ. Corn smut dikaryon in culture. Nature New Biol 1971; 23: 19-20.

[24]   Banuett F and Herskowitz I. Different *a* alleles of *Ustilago maydis* are necessary for maintenance of filamentous growth but not for meiosis. Proc Natl Acad Sci USA 1989; 86: 5878-5882.

[25]   Puhalla JE. Compatibility reactions on solid medium and interstrain inhibition in *Ustilago maydis.* Genetics 1968; 60: 461-474.

[26]   Rowell JB, DeVay JE. Genetics of *Ustilago zeae* in relation to basic problems of its pathogenicity. Phytopathology 1954; 44: 356-362.

[27]   Kronstad JW and Leong SA. Isolation of two alleles of the *b* locus of *Ustilago maydis.* Proc Natl Acad Sci USA 1989; 86: 978-982.

[28]   Wangemann-Bude M and Schauz K. Intraspecific hybridization of *Ustilago maydis* haploids with compatible and incompatible mating type by electrofusion and genetic analysis of fusion products. Exp Mycol 1991; 15: 159-166.

[29]   García-Pedrajas MD, Nadal M, Bölker M, Gold SE, Perlin MH. Sending mixed signals: Redundancy *vs.* uniqueness of signaling components in the plant pathogen *Ustilago maydis.* Fungal Genet Biol 2008; 45: S22-S30.

[30]   León-Ramírez CG, Cabrera-Ponce JL, Martínez-Espinoza AD, Herrera-Estrella L, Méndez L, Reynaga-Peña CG and Ruiz-Herrera J. Infection of alternative host plant species by *Ustilago maydis.* New Phytol 2004; 164: 337-346.

[31]   Méndez-Morán L, Reynaga-Peña CG, Springer PS and Ruiz-Herrera J. *Ustilago maydis* infection of the non-natural host *Arabidopsis thaliana.* Phytopathology 2005; 95: 480-488.

[32]   Flor-Parra I, Vranes M, Kämper J, Perez-Martin J. Biz1, a zinc finger protein required for plant invasion by *Ustilago maydis,* regulates the levels of a mitotic cyclin. Plant Cell 2006; 18: 2369-2387.

[33]   Brachmann A, Schirawski J, Müller P and Kahmann R. An unusual MAP kinase is required for efficient penetration of the plant surface by *Ustilago maydis.* EMBO J 2003; 22: 2199-2210.

[34]   Schirawski J, Böhnert HU, Steinberg G, Snetselaar K, Adamikowa L and Kahmann R. Endoplasmic reticulum glucosidase II is required for pathogenicity of *Ustilago maydis.* Plant Cell 2005; 17: 3532-3543.

[35]   Doehlemann G, Wahl R, Vranes M, Vries RP, Kämper J, Kahmann R. Establishment of compatibility in the *Ustilago maydis*/maize pathosystem. J Plant Physiol 2008; 165: 29-40.

[36]   Molina L, Kahmann R. An *Ustilago maydis* gene involved in H_2O_2 detoxification is required for virulence. Plant Cell 2007; 19: 2293-2309.

[37]   Reineke G, Heinze B, Schirawski J, Buettner H, Kahmann R, Basse CW. Indole-3-acetic acid (IAA) biosynthesis in the smut fungus *Ustilago maydis* and its relevance for increased IAA levels in infected tissue and host tumour formation. Mol. Plant Pathol 2008; 9: 339-355.

[38]   Rodríguez-Kessler MA, Ruiz OA, Maiale S, Ruiz-Herrera J and Jiménez-Bremont JF. Polyamine metabolism in maize tumors induced by *Ustilago maydis.* Plant Physiol Biochem 2008; 46: 805-814.

[39] Skibbe DS, Doehlemann G, Fernandes J and Walbot V. Maize tumors caused by *Ustilago maydis* require organ-specific genes in host and pathogen. Science 2010; 328: 89-92.

[40] Ruiz-Herrera J, León-Ramírez C, Cabrera-Ponce JL, Martínez-Espinoza AD and Herrera-Estrella L. Completion of the sexual cycle and demonstration of genetic recombination of *Ustilago maydis in vitro*. Mol Gen Genet 1999; 262: 468-472.

[41] Gold S, Duncan G, Barret K and Kronstad J. cAMP regulates morphogenesis in the fungal pathogen *Ustilago maydis*. Gene Develop 1994; 8: 2805-2816.

[42] Klose J, Moniz de Sa MM, and Kronstad JW. Lipid-induced filamentous growth in *Ustilago maydis*. Mol Microbiol 2004; 52: 823–835.

[43] Ruiz-Herrera J, León CG, Guevara-Olvera L and Cárabez-Trejo A. Yeast-mycelial dimorphism of haploid and diploid strains of *Ustilago maydis*. Microbiology 1995; 141: 695-703.

[44] Sánchez-Martínez C, Pérez-Martin J. Dimorphism in fungal pathogens: *Candida albicans* and *Ustilago maydis*-similar inputs, different outputs. Curr Opinion Microbiol 2001; 4: 214-221.

[45] Martínez-Espinoza AD, Garcia-Pedrajas MD and Gold SE. The Ustilaginales as plant pests and model systems. Fungal Genet Biol 2002; 35: 1-20

[46] Banuett F. Pathogenic development in *Ustilago maydis*: a progression of morphological transitions that results in tumor formation and teliospore production. In: Osiewacz, H.D. (Ed.), Molecular Biology of Fungal Development. Marcel Dekker, New York, Basel 2002; pp 349–398.

[47] Dürrenberger F, Wong K and Kronstad JW. Identification of a cAMP-dependent protein kinase catalytic subunit required for virulence and morphogenesis in *Ustilago maydis*. Proc Natl Acad Sci USA 1998; 95: 5684-5689.

[48] Martínez-Espinoza AD, Ruiz-Herrera J, León-Ramírez CG and Gold SE. MAP kinase and cAMP signaling pathways modulate the pH-induced yeast-to-mycelium transition in the corn smut fungus *Ustilago maydis*. Curr Microbiol 2004; 49: 274-281.

[49] Klose J and Kronstad JW. The multifunctional "-oxidation enzyme is required for full symptom development by the biotrophic maize pathogen *Ustilago maydis*. Eukaryot Cell 2006; 5: 2047-2061.

[50] Ruiz-Herrera J. The role of polyamines in fungal cell differentiation. Arch Med Res 1993; 24: 263-265.

[51] Ruiz-Herrera J. Polyamines, DNA methylation, and fungal differentiation. Crit Rev Microbiol 1994; 20: 143-150.

[52] Guevara-Olvera L, Xoconostle-Cázares B and Ruiz-Herrera J. Cloning and disruption of the ornithine decarboxylase gene of *Ustilago maydis*. Evidence for role of polyamines in the dimorphic transition. Microbiology 1997; 143: 2237-2245.

[53] Valdés-Santiago L, Cervantes-Chávez JA and Ruiz-Herrera J. *Ustilago maydis* spermidine synthase is encoded by a chimeric gene, required for morphogenesis, and indispensable for survival in the host. FEMS Yeast Res 2009; 9: 923-935.

[54] León-Ramírez CG, Valdés-Santiago L, Campos-Góngora E, Ortiz-Castellanos L, Aréchiga-Carvajal ET and Ruiz-Herrera J. A molecular probe for Basidiomycota: the spermidine synthase-saccharopine dehydrogenase chimeric gene. FEMS Microbiol Lett 2010; 312: 77-83.

[55] Reyna-López G, Simpson J and Ruiz-Herrera J. Differences in DNA methylation patterns are detectable during the dimorphic transition of fungi by amplification of restriction polymorphims. Mol Gen Genet 1997; 6: 703-710.

[56] Antequera F, Tamame M, Villanueva JR and Santos T. DNA methylation in the fungi. J Biol Chem 1985; 259: 8033-8036.

[57] Magill JM, and Magill CW. DNA methylation in fungi. Develop Genet 1989; 10: 63-69.

[58] Binz T, D'Mello N, Horgen PA. A comparison of DNA methylation levels in selected isolates of higher fungi. Mycologia 1998; 90: 785-790.

[59] Reyna-López GE and Ruiz-Herrera J. Specificity of DNA methylation changes during fungal dimorphism and its relationship to polyamines. Curr Microbiol. 2004; 40: 118-123.

[60] Cano C, Herrera-Estrella L and Ruiz-Herrera J. DNA methylation and polyamines in the regulation of development of the fungus *Mucor rouxii*. J Bacteriol 1988; 170: 5946-5948.

[61] Cano-Canchola C, Sosa L, Fonzi W, Sypherd P and Ruiz-Herrera J. Developmental regulation of *CUP* gene expression through DNA methylation in *Mucor* spp. J Bacteriol 1992; 174: 362-366.

[62] Ruiz-Herrera J, Ruiz-Medrano R and Domínguez A. Selective inhibition of cytosine-DNA methylases by polyamines. FEBS Lett 1995; 357: 192-196.

[63] Peñalva MA and Arst HNJr. Regulation of gene expression by ambient pH in filamentous fungi and yeast. Microbiol Mol Biol Rev 2002; 66: 426-446.

[64]   Peñalva MA, and Arst HNJr. Recent advances in the characterization of ambient pH regulation of gene expression in filamentous fungi and yeasts. Annu Rev Microbiol 2004; 58: 425-451.

[65]   Peñalva MA, Tilburn J, Bignell E and Arst HNJr. Ambient pH gene regulation in fungi: making connections. Rev Trends Microbiol 2008; 16: 291-300.

[66]   Aréchiga-Carvajal ET and Ruiz-Herrera J. The Rim101/PacC homologue from the Basidiomycete *Ustilago maydis* is functional in multiple pH-sensitive phenomena. Eukaryot Cell 2005; 4: 999-1008.

[67]   Cervantes-Chávez JA, Ortiz-Castellanos L, Tejeda-Sartorius M, Gold S and Ruiz-Herrera J. Functional analysis of the pH responsive pathway Pal/Rim in the phytopathogenic basidiomycete *Ustilago maydis*. Fungal Genet Biol 2010; 47: 446-457.

[68]   Ramon AM, Porta A and Fonzi WA. Effect of enviromental pH on morphological development of *Candida albicans* is mediated *via* the pacC-related transcription factor encoded by PRR2. J. Bacteriol 1999; 181: 7524-7530.

[69]   Davis D, Wilson RB, and Mitchell AP. Rim101-dependent and -independent pathways govern pH responses in *Candida albicans*. Mol Cell Biol 2000; 20: 971-978.

[70]   Ruiz-Herrera J, León CG, Carabez-Trejo A and Reyes-Salinas E. Structure and chemical composition of the cell walls from the haploid yeast and mycelial forms of *Ustilago maydis*. Fungal Genet Biol 1996; 20: 133-142.

[71]   Gold SE and Kronstad JW. Disruption of two genes for chitin synthase in the phytopathogenic fungus *Ustilago maydis*. Mol Microbiol 1994; 11: 897-902.

[72]   Xoconostle-Cázares B, León-Ramírez C and Ruiz-Herrera J. Two chitin synthase genes from *Ustilago maydis*. Microbiology 1996; 142: 377-387.

[73]   Xoconostle-Cázares B, Specht CA, Robbins PW, Liu Y, León C and Ruiz-Herrera J. *Umchs5*, a gene coding for a class IV chitin synthase in *Ustilago maydis*. Fungal Genet Biol 1997; 22: 199-208.

[74]   Garcerá-Teruel A, Xoconostle-Cázares B, Rosas-Quijano R, Ortiz L, León-Ramírez C, Specht CA, Sentandreu R and Ruiz-Herrera J. Loss of virulence in *Ustilago maydis* by *Umchs6* gene disruption. Res Microbiol 2004; 155: 87-97.

[75]   Weber I, Assmann D, Thines E and Steinberg G. Polar localizing class V myosin chitin synthases are essential during early plant infection in the plant pathogenic fungus *Ustilago maydis*. Plant Cell 2006; 18: 225-242.

[76]   Bartnicki-Garcia S, Lippman E. Fungal morphogenesis: cell wall construction in *Mucor rouxii*. Science 1969; 165: 302-304.

[77]   Girbardt M. Lebendbeobachtungen on *Polystictus versicolor* (L.) Flor 1955; 142: 540-563.

[78]   Girbardt M. Der Spitzenkorper von *Polystictus versicolor* (L.) Planta 1957; 50: 47-59.

[79]   Grove SN, Bracker CE. Protoplasmic organization of hyphal tips among fungi: vesicles and Spitzenkorper. J Bacteriol 1970; 104: 989-1009.

[80]   Banuett F, Quintanilla RHJr, Reynaga-Peña CG. The machinery for cell polarity, cell morphogenesis, and the cytoskeleton in the Basidiomycete fungus *Ustilago maydis*- A survey of the genome sequence. Fungal Genet Biol 2008; 45: S3-S14.

INDEX

Tup 37, 39, 61, 62
TUP1, 37, 39, 61, 62
Turkey, 88
Tutipotent, 5
Tweens, 110

U

U. maydis, 8, 9, 10, 11, 40, 60, 61, 62, 105, 106, 107, 108, 109, 110, 111, 112, 113
Ubiquinone oxidoreductase, 23
UDP_glucose, 16, 47
Ulcerated, 68, 72
Ulcers, 70, 71
Ume6, 38, 39
Uncastrated, 75
Unfolding, 52
Unsaturated, 9, 19, 20, 50, 51, 58, 88
Upregulated, 11, 19, 23, 40, 108
URA3, 36
Urine, 88
Uruguay, 74, 75
USA, 46, 68, 74, 75, 78
Ustilaginales, 105
Ustilago hordei, 105
Ustilago maydis, 4, 10, 36, 59, 61, 105, 106, 107
Ustilago nigra, 105
Ustilago nuda, 105

V

Vaccine, 51, 53, 54
Vaginal, 35, 40, 88
Valvular heart disease, 71
V-ATPase, 52, 53
Vectorial, 7, 8
Vectorial phenomenon, 7
Vectorial processes, 7, 8
Vectors, 26
Vegetative, 61, 87, 96, 97
Vein, 77
Venezuela, 16, 24, 25
Verapamil, 93
Vesicle supply centre, 95
Vesicles, 12, 95, 113
Vessels, 36, 70, 78
Veterinary, 72, 76
Viability, 62
Virulence, 3, 4, 10, 20, 25, 35, 36, 38, 39, 40, 41, 46, 47, 48, 49, 51, 52, 53, 54, 59, 62, 67, 68, 69, 70, 76, 77, 78, 80, 105, 108, 109, 110, 113
Virulence determinants, 46, 51
Virulence factors, 25, 67, 68, 70, 76, 77, 80
Virulent, 4, 17, 35, 36, 37, 38, 40, 47, 49, 50, 51, 52, 53, 54, 59, 107, 108, 109, 113
Virulent stage, 4

W

WAP1, 39
Wastewater, 58

www.ingramcontent.com/pod-product-compliance
Lightning Source LLC
Chambersburg PA
CBHW041713210326
41598CB00007B/633